The One Minute Cat Ma

Kac Young PhI

Sixty seconds to feline

Illustrations by Lara Mesanza Burke

Hubble & Hattie

The Hubble & Hattie imprint was launched in 2009 and is named in memory of two very special Westies owned by Veloce's proprietors. Since the first book, many more have been added to the list, all with the same underlying objective: to be of real benefit to the species they cover, at the same time promoting compassion, understanding and respect between all animals (including human ones!)

Hubble & Hattie is the home of a range of books that cover all-things animal, produced to the same high quality of content and presentation as our motoring books, and offering the same great value for money.

More great Hubble & Hattie books!

Among the Wolves: Memoirs of a wolf handler (Shelbourne)
Animal Grief: How animals mourn (Alderton)
Babies, kids and dogs – creating a safe and harmonious relationship (Fallon & Davenport)
Because this is our home ... the story of a cat's progress (Bowes)
Bonds – Capturing the special relationship that dogs share with their people (Cukuraite & Pais)
Camper vans, ex-pats & Spanish Hounds: from road trip to rescue – the strays of Spain (Coates & Morris)
Canine aggression – how kindness and compassion saved Calgacus (McLennan)
Cat and Dog Health, The Complete Book of (Hansen)
Cat Speak: recognising & understanding behaviour (Rauth-Widmann)
Charlie – The dog who came in from the wild (Tenzin-Dolma)
Clever dog! Life lessons from the world's most successful animal (O'Meara)
Complete Dog Massage Manual, The – Gentle Dog Care (Robertson)
Confessions of a veterinary nurse: paws, claws and puppy dog tails (Ison)
Detector Dog – A Talking Dogs Scentwork Manual (Mackinnon)
Dieting with my dog: one busy life, two full figures ... and unconditional love (Frezon)
Dinner with Rover: delicious, nutritious meals for you and your dog to share (Paton-Ayre)
Dog Cookies: healthy, allergen-free treat recipes for your dog (Schöps)
Dog-friendly gardening: creating a safe haven for you and your dog (Bush)
Dog Games – stimulating play to entertain your dog and you (Blenski)
Dog Relax – relaxed dogs, relaxed owners (Pilguj)
Dog Speak: recognising & understanding behaviour (Blenski)
Dogs just wanna have Fun! Picture this: dogs at play (Murphy)
Dogs on Wheels: travelling with your canine companion (Mort)
Emergency First Aid for dogs: at home and away Revised Edition (Bucksch)
Exercising your puppy: a gentle & natural approach – Gentle Dog Care (Robertson & Pope)
For the love of Scout: promises to a small dog (Ison)
Fun and Games for Cats (Seidl)
Gods, ghosts, and black dogs – the fascinating folklore and mythology of dogs (Coren)
Helping minds meet – skills for a better life with your dog (Zulch & Mills)
Home alone – and happy! Essential life skills for preventing separation anxiety in dogs and puppies (Mallatratt)
Know Your Dog – The guide to a beautiful relationship (Birmelin)
Letting in the dog: opening hearts and minds to a deeper understanding (Blocker)
Life skills for puppies – laying the foundation for a loving, lasting relationship (Zuch & Mills)
Lily: One in a million! A miracle of survival (Hamilton)
Living with an Older Dog – Gentle Dog Care (Alderton & Hall)
Miaow! Cats really are nicer than people! (Moore)
Mike&Scrabble – A guide to training your new Human (Dicks & Scrabble)
Mike&Scrabble Too – Further tips on training your Human (Dicks & Scrabble)
My cat has arthritis – but lives life to the full! (Carrick)
My dog has arthritis – but lives life to the full! (Carrick)
My dog has cruciate ligament injury – but lives life to the full! (Haüsler & Friedrich)
My dog has epilepsy – but lives life to the full! (Carrick)
My dog has hip dysplasia – but lives life to the full! (Haüsler & Friedrich)
My dog is blind – but lives life to the full! (Horsky)
My dog is deaf – but lives life to the full! (Willms)
My Dog, my Friend: heart-warming tales of canine companionship from celebrities and other extraordinary people (Gordon)
Office dogs: The Manual (Rousseau)
One Minute Cat Manager: sixty seconds to feline Shangri-la (Young)
Ollie and Nina and ... daft doggy doings! (Sullivan)
No walks? No worries! Maintaining wellbeing in dogs on restricted exercise (Ryan & Zulch)
Partners – Everyday working dogs being heroes every day (Walton)
Puppy called Wolfie – a passion for free will teaching (Gregory)
Smellorama – nose games for dogs (Theby)
Supposedly enlightened person's guide to raising a dog (Young & Tenzin-Dolma)
Swim to recovery: canine hydrotherapy healing – Gentle Dog Care (Wong)
Tale of two horses – a passion for free will teaching (Gregory)
Tara – the terrier who sailed around the world (Forrester)
Truth about Wolves and Dogs, The: dispelling the myths of dog training (Shelbourne)
Unleashing the healing power of animals: True stories about therapy animals – and what they do for us (Preece-Kelly)
Waggy Tails & Wheelchairs (Epp)
Walking the dog: motorway walks for drivers & dogs revised edition (Rees)
When man meets dog – what a difference a dog makes (Blazina)
Wildlife photography from the edge (Williams)
Winston ... the dog who changed my life (Klute)
Wonderful walks from dog-friendly campsites throughout the UK (Chelmicka)
Worzel Wooface: For the love of Worzel (Pickles)
Worzel Wooface: The quite very actual adventures of (Pickles)
Worzel Wooface: The quite very actual Terribibble Twos (Pickles)
Worzel Wooface: Three quite very actual cheers for (Pickles)
You and Your Border Terrier – The Essential Guide (Alderton)
You and Your Cockapoo – The Essential Guide (Alderton)
Your dog and you – understanding the canine psyche (Garratt)

Hubble & Hattie Kids!

Fierce Grey Mouse (Bourgonje)
Indigo Warrios: The Adventure Begins! (Moore)
Lucky, Lucky Leaf, The: A Horace & Nim story (Bourgonje & Hoskins)
Little house that didn't have a home, The (Sullivan & Burke)
Lily and the Little Lost Doggie, The Adventures of (Hamilton)
Wandering Wildebeest, The (Coleman & Slater)
Worzel goes for a walk! Will you come too? (Pickles & Bourgonje)
Worzel says hello! Will you be my friend? (Pickles & Bourgonje)

www.hubbleandhattie.com

Images courtesy Lara Mesanza Burke

First published in April 2019 by Veloce Publishing Limited, Veloce House, Parkway Farm Business Park, Middle Farm Way, Poundbury, Dorchester, Dorset, DT1 3AR, England. Tel 01305 260068/fax 01305 250479/e-mail info@hubbleandhattie.com/ web www.hubbleandhattie.com. ISBN: 978-1-787113-73-2 UPC: 6-36847-01373-8. Readers with ideas for books about animals, or animal-related topics, are invited to write to the editorial director of Veloce Publishing at the above address. British Library Cataloguing in Publication Data - A catalogue record for this book is available from the British Library. Typesetting, design and page make-up all by Veloce Publishing Ltd on Apple Mac. Printed and bound in India by Replika Press PTY

Contents

Acknowledgments

I owe a huge debt of gratitude to the following miracles in my life –

Marlene Morris, whose creativity and inspiration fill my life, who generously penned many of the wonderful bedtime stories for cats, who let me run every wacky cat-related idea by her, and then helped me try them on the cats.

Lisa Hagan, who is a cat lover and a spectacular agent; Jude Brooks, a brilliant publisher and animal lover; Lara Mesanza Burke, the talented artist who brought life to the stories; Lisa Tenzin-Dolma, my dear friend, cheerleader and extraordinary writer and inspiration; David Congalton, my friend and gifted author, screenwriter and radio show host; Lyla Oliver, longtime friend and successful television writer with a killer wit and a liberal heart; Dr Colleen Kennedy, my oldest and dearest friend, and mother to my godcat, Ziggy.

Pamela Ventura, the best sister-in-law you could ever have; talented, funny and a gifted person and television writer; Beth Wareham, extraordinary friend, writer, publisher and advocate of all that's right and just; Tracy Abbott Cook, another brilliant writer, funny-woman and trusted, cherished friend; Donna Wells, attorney, mother of my goddaughter and always willing and able to support my madcap ideas; Patrick Morris, stepson: there is no end to his talent and heart; Peggy Jones, another treasured friend and sister who helped me sneak 'Dot,' an adopted kitten, onboard a flight decades back, and who also helped me find and rescue the beloved Jabez.

Donna Smith, movie producer and incredible friend who started it all forty-one years ago with Lucy the rescue kitten; Sarah Hartwell, for her generous gift of the cat glossary and PETA UK; HART, Cambria, CA;

Sam Ratcliffe and Laura Nativo, who recommended Daniel Quagliozzi of Go Cat Go to write the Foreword, and all the magnificent people in the world who save, rescue, foster and adopt the magnificent creatures we call cats ...

Dedication

To Miss Lucy,
with all my love

Foreword

by Daniel 'DQ' Quagliozzi

Unraveling the mysteries of the feline message has always fascinated me. I think people misunderstand just how complex feline emotions and body language can be. You literally have to stop yourself from projecting your own feelings on to your cat. Never take it personally! It's not always about YOU!

Many of the consulting clients I have want to just fit a cat into their routine and ignore how the cat is feeling. I work with them to establish a one-on-one bond with their, cat and teach them how they can make a few shifts in their lifescape to accommodate the needs of their feline companion. Everybody wins.

The One Minute Cat Manager gets straight to the point. I love how direct it is and how useful the information is for first-time cat owners (newbies), and seasoned cat owners alike. This book lays it out for you. It describes a cat's body language, personality, and some signals that can easily be missed. It doesn't take long – just a minute or two – to figure out what a cat needs, and what results is a more harmonious relationship between human and feline.

Understanding your cat is the number one task for cat owners. 90% of my consultations focus on the human dilemma, but understanding that making a few adjustments in consideration of the cat's nature and habits can immediately result in a happier home. I always like to encourage my clients to 'live in the Meow,' and this book does that. It helps the human being understand the world of the cat, quickly.

In the book there are a lot of excellent safety precautions, instructions for walking a cat on a harness, and tips on keeping your cat healthy and happy. It's a fun read, and I laughed out loud a few times at the stories and antics of the featured cats.

I especially recommend spending time with your cat. Whether it's telling your cat all about your day or reading them a bedtime story from chapter fifteen of the book, it's all about the sound of your voice to your cat's ears, and the soothing nature of your presence.

There are a lot of things you can't control about a cat, but there are many things you can. *The One Minute Cat Manager* helps us understand that the cat isn't 'wrong' or 'bad,' he's just living according to his own nature, and if we can slow down a tad we can learn more about his world; in a minute.

One of the greatest gifts a person can have is a loving bond and relationship with a cat. If I can help people understand more about the internal workings of their cat, there will be a higher cat adoption retention percentage and better symbiosis between cat and owner.

A high number of cats are recycled each year, and if more people would read this book they would become better educated about cats, and thereby increase the success rate of cat adoptions worldwide. I've seen the inside of cat shelters firsthand, and I know that knowledge is power.

Written in a clear and authentic voice with the heart of a cat lover and an experienced cat parent, everyone thinking about adopting a cat should read this book, and anyone who has a cat should read it, too.

I will absolutely recommend this book to my clients as I coach them into being a caring and responsible feline parent, so they can have years of joy and love with their equally happy cat.

Daniel 'DQ' Quagliozzi
San Francisco, California

Daniel shares his private time with his mischievous, stubby-legged Siamese cat, Cubby.

Originally from New Jersey, Daniel now resides in San Francisco, and spent a decade working with the San Francisco SPCA as a specialist in cat behavior, intake and adoption. Daniel brings a breadth of knowledge and a compassionate approach to his consulting clients, sharing what he has learned on the frontlines of animal welfare. He is an expert on cat behavior and has appeared on radio and television talking about feline body language, aggression, fear, compassion, and the human-animal bond. His company is Go, Cat, Go (https://gocatgosf.com/). Daniel has been featured in the *San Francisco Chronicle*, *San Francisco Magazine*, and *SFWEEKLY*.

You can hear him dishing out behavior advice on KGO Radio as a regular guest on the Maureen Langan show, and as a cat expert on KOIT Radio.

Go, Cat, Go is a Bay Area A List winner of Best Pet Training in 2014, 2015, 2016 & 2017.

Introduction

How did a kitty come into your life? Were your heartstrings tugged when you saw that little ball of fluff meowing and needing a home? Did you find him injured and abandoned at the side of the road? Perhaps she was in a shelter just waiting for you to rescue her ... or maybe you paid a pretty penny to purchase an exotic breed? Perhaps you were being a good neighbor and took in kitty as a gesture of friendship?

If you share your life with a cat, then you probably fit one of these familiar scenarios: after all, it's unlikely you got him online from Amazon or won her at the State Fair. We are all chosen people when it comes to cats.

We might think, for a brief moment, that we are the owner of the cat because we adopted him, signed a few papers, gave him a name and erroneously assumed we are his 'parents' because we are in charge, and they can't vote or drive.

What has really happened, though, is that we have succumbed to the cat's charms and wiles, falling under this spell in a very short time after being selected by the cat concerned. I was among those who felt I was giving kitty a second chance: a warm place to sleep and food to survive. I suppose I even expected a soupçon of gratitude from her, but what I got was an enormous lesson in unconditional love and a lifetime of entertainment and adorableness. The cat is *always* in charge ... but I had to learn that over time.

I've had cats in my life for over 40 years and can't imagine living without them: their dear little faces following me around and ordering me about. I love them so much. Which is why, in all those decades of experience, I have learned all about cats from vets, cat experts, cat clubs, cat magazines, and a whole host of other sources. So much so, that my friends always call me for cat-related advice before going online to research a problem.

Chances are, I have had a cat with the issues they describe, and resolved them with a veterinarian, an holistic practitioner, or all by myself. I had to learn what was a DIY situation, and what needed professional attention. My twenty-plus felines gave me many situations to explore, solve and heal. It is because of them that I became savvy in the ways of a cat.

The most incredible spirit of strength and determination I ever met came in the form of a coal-black kitten, who triumphed over all humans I have met to date. Her name was Lucy, and she was born inside the wall of my friend's house, when her frightened, feral momma gave birth between the garbage disposal and the plumbing pipes in the supporting wall between the exterior and the kitchen sink. My friend, Donna, heard the cries of the new kittens, but didn't know how to reach them, so she did what any red-blooded American girl would do: she phoned the fire department. Of course, they could not dispatch an entire truck and crew to rescue the cat and kittens, but a generous firefighter told her he would come by after his shift and do what he could. He kept his word.

He crafted a little box, hoisted by a pulley rig fashioned out of kitchen implements, and lowered the box like a little elevator between the spaces in the wall, rescuing the kittens from their plasterboard prison. The momma, terrified by the goings-on, split. Donna found herself with four kittens that were only a day old, so quickly phoned a local vet for advice, after placing the kittens in a shoebox with washcloths to keep them warm.

Just as the hero firefighter was getting ready to leave, Donna heard one more tiny voice coming from the wall. It was pitch black inside the wall, and so was the miniscule fifth kitten. Donna and the fireman used a long-handled wooden spoon to scoop her up and haul her up and out of the wall, using the makeshift elevator box. This little pistol was fighting for her life at the ripe old age of 24 hours. She was winter-sky black. Donna named her Lucy.

At that time, Donna was working on a show called *Circus of the Stars*. The location for shooting the CBS extravaganza was Caesar's Palace in Las Vegas, and Donna had no choice but to take the kittens with her on the trip. Very quickly, she made friends with the hotel housekeepers who were put 'on guard' to monitor the kittens. Donna raced back and forth from the shoot to her room to bottle feed the kittens eight times a day (and night). By day, Donna was coordinating film and television stars, lion acts, elephant acts, and trapeze acts. By night, she was feeding kittens with tiny bottles and droppers.

Donna Smith and her husband, Gordon, had been saving feral cats for decades. They humanely trapped them, brought them inside into their 'cat room' (spare bedroom), took them in for medical care, and had them spayed or neutered before finding them forever homes. They were amazing. This kitten care was a gesture of love and never a burden. Donna was an old-hand at kitten rescue and placement.

Lucy and her siblings lived for two weeks at Caesar's palace before traveling back to Los Angeles by car. They were growing fast, had opened their eyes, and were doing well for wall-born cats. And they were getting harder to handle: they had long-since outgrown their shoebox, and had graduated to a milk crate.

Back home, they had full reign in the cat room, and Donna was enjoying their kitten pranks.

A few weeks later Donna came by to visit me. I had adopted one of her previously-trapped cats, and she was a Siamese beauty. I knew nothing about cats at the time, but Donna quizzed me and gave me lessons, and determined that I would be a fit mother for the 6-month-old Siamese who needed a home. (I mean she actually gave me a written test!) She was serious about cat parents.

I was on my way out to dinner when Donna knocked on the door. She was toting a box containing several kittens, and, thoughtfully, she wanted me to have the 'pick of the litter' before she found the rest homes. I told her how nice that was but I didn't want another cat; especially a kitten, though they were very, very cute.

Lucy was the first one out of the box, and proceeded to climb into my lap ... where she stayed for the rest of the evening. Summer, the Siamese, already self-proclaimed queen of the house, took one look at Lucy and instantly turned into mother of the year. Lucy and Summer bonded right away. I was smitten with this kitten and we never got to dinner; we had too much fun playing with the cats and forgot to eat. Oh, well.

Lucy lived to be twenty-four, and is the inspiration for *The One Minute Cat Manager*. She taught me everything I know about cats, and then some. She had an opinion about everything and let me stumble along until I became an adequate cat-parent. Which is probably the reason she hung on for twenty-four years. (There will be more about Lucy in the coming chapters, but I thought you should know, right up front, that she is the main reason for this book, and the system of *The One Minute Cat Manager*.)

Kac Young

1 The experiment

When I was hired for my first corporate job I didn't know a thing about corporate politics, reporting to a specific chain of command, or juggling a workload equal to a space shuttle launch. Previously, I had worked in freelance television where the hierarchy was made clear by the power certain positions held. Stars, producers, directors and writers topped the list; then came the assistants, and finally the crew and technical people. We called the top creative people Above the Line and the crew and support staff Below the Line. I never understood what the 'line' was, but it was clear that your position and pay grade was either above it or below this ... whatever it was.

Joining a big corporation was a completely different animal and a fresh learning curve. Bosses had titles, and were lined up in order of the 'Chiefs' (CFO, CEO, COO, CMO, CTO) or President, Vice President, and even those titles had qualifiers such as Senior or First, Second or Third. Below the VPs were Directors, Assistant Directors, and Managers. Some of these titles were also designated by other qualifiers or divisions. It all gave me a migraine. Coming from my showbusiness background, the only Chairman I knew was Frank Sinatra ... everyone else was an imposter.

I was called to 'the principal's office' several times in my first few months. I wasn't mastering the hang of the corporate hierarchy. It never occurred to me that if I wanted a meeting with the Chairman or the CFOOL, that I couldn't simply email them directly and ask for it. No, no. The proper way to 'ask permission' was to go up through my boss, and request that he set the meeting on my behalf. What a waste of time!

One day my boss called me into his office to tell me that he'd never had as many calls from his bosses in just one day as when I emailed

The Chairman. My boss was summoned out of a lunch, harassed by his superiors, and ordered to find out what was going on. I told him that I had simply invited our Chairman to an Award Show we were shooting on the lot because I thought he might enjoy mingling with all of the stars we were hosting. I even got him a seat at the head table.

My boss called *his* boss there and then, and asked: "Are we at war; is something on fire; is anyone bleeding?" "No." "We're good, then." he replied. This pompous, ego-tromping incident led me to expand my education in corporate protocol. I bought *The One Minute Manager* by Ken Blanchard and Spencer Johnson, which gave me many insights and specific direction into becoming an effective corporate manager. My learning curve was steep.

I was mostly attracted to the one minute part of the book since anything that was succinct and to the point got my attention. In television everything is pretty much thirty second sound bites and instant gratification; one minute seemed like a luxurious amount of time. The lessons and insights I gained stayed with me and crossed over into other areas of my life. It appeared that most of what you have to accomplish and manage with people can happen in a minute. One minute of interaction can turn out to be a boon, or a disaster, depending on how you fashion it. You can also end your career in a minute if something goes wrong, as I had been on the brink of discovering.

One day, in the midst of a horrendously busy schedule, while rushing to catch yet another airplane for an endless series of meetings, I caught the look on Lucy's face. She looked extremely sad, forlorn, and if about to burst into kitty tears.

Lucy had seen the suitcases, jumped inside, and knew I was going away – again. I couldn't change the reality of the trip but a brilliant flash of the obvious swept across my brain. What if I tried *The One Minute Manager* technique on her?

Before ...

For one complete minute, I held her, petted her, whispered sweet everythings into her little ear, and told her that I loved her as many times as it's possible to say in 60 seconds. I promised Lucy we would have many more minutes together when I got home.

It seemed to work; it was a life-changing minute for us. I remember boarding the plane and sitting in my seat in readiness for the cross-country flight, proudly thinking I had really connected with my little feline darling. If Lucy felt as good as I did, this would hold us both until I returned home.

Fresh on the heels of my new discovery, I began applying the One Minute philosophy to everything in my pet's life. Not only did we accomplish the affectionate loving-on-one-another process, but it also worked for health checks, combing rituals, behavior modification, feeding and snuggling sessions. In just One Minute we were able to accomplish loads, and I've not felt guilty since. I began passing on *The One Minute Cat Manager* to friends and co-workers. It caught on like wildfire.

When we find ourselves too caught up in our own lives to pay adequate attention to our pets, or when we are preoccupied, distracted or just in a bad mood from what's been going on at work that day, we can remember to take one minute for them. Instead of bursting in the door and heading for the fridge, bar, or television, we can choose to greet them and play with them for one minute, since they've spent the entire day without us. That one minute means a lot.

Later on, we can spend more time with them as we unwind from the day and decompress. Naturally, we want to spend as much personal time with

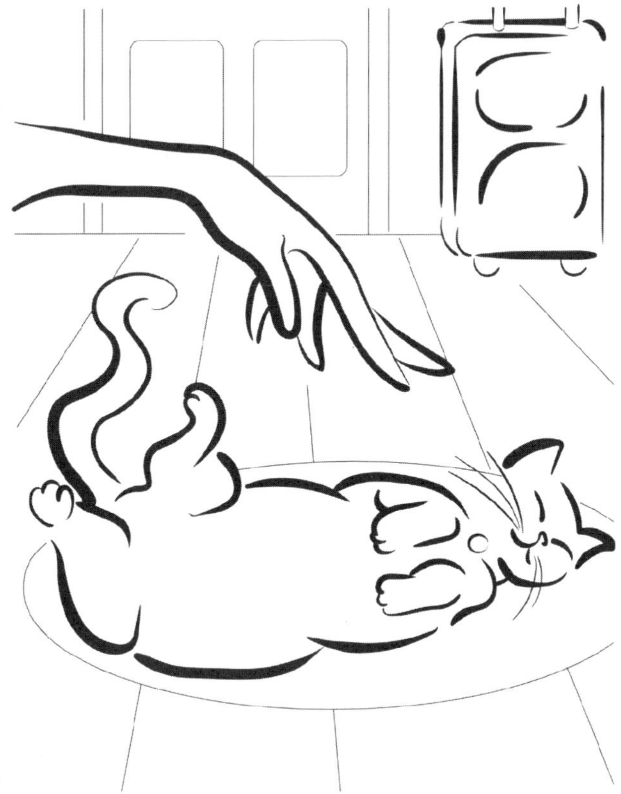

After ...

our furry friends as we can but, in a pinch, one minute of focused attention helps us create a better relationship with our felines.

When you can't spend a lot of time with your cat, for one reason or another, try this system and give your cat a dedicated minute of your undivided attention. It may be a band aid on the problem, but it truly helps in the short term. Both you and your cat will feel the benefit from one solid minute of deep communication.

Because this system worked very well, I became the resident cat expert on my floor at work. I gave away copies of *The One Minute Manager* like candy, and everyone was excited about using this system on their pets. I was able to adopt a few more rescue cats because, by using this system, I could meet all their needs. I love my cats so much, and I think that maybe the example of our happy feline home inspired several of my friends to adopt a cat, and use this one minute system. Everybody wins.

This book is a collection of what I have learned over the years, what others have told me about their own experience with the system, and what I have seen work. I pass it along to you, so you can enjoy your cat(s), and practice a system that rewards you and your feline(s) for the effort.

You can master *The One Minute Cat Manager* system easily, just as I did two decades ago: the technique helps you to focus on specific needs when you don't think you have a second to spare, and creates a bond of focus and attention with your cat, even when you're having to race out the door. Everybody can always find a minute.

One Minute Cat Manager summary

- You can accomplish most interactions in one minute or under
- Many of life's most important connections occur in a minute or less
- Taking great care of your cat can be broken down into one minute increments
- These may become the best minutes you ever spend in your life

2 How to read your cat

When you decided to enrich your life and give a forever home to a cat, you bonded with your precious feline, and now you are eager to get to know him better. You wonder: How does she feel? What does he think? Does she love me? How will I know? These are very human questions and can be answered through the basic language of 'Cat,' a language full of feelings, positions, movements, glances, and sounds. You'll learn to converse in 'Cat' in no time ... here's your first application of *The One Minute Cat Manager.*

Just as cats have different tastes in foods, they also have different emotional needs. Every cat needs attention physically, emotionally, and spiritually. They communicate their wants and needs through a language of sounds, movement, and positions. It's like the expressive Italian who talks with his hands, only cats don't use their paws: they simply express themselves like a dancer, using body and personality to communicate feelings. This is the secret to how their language works.

We know from experience that some cats are very sociable, and enjoy being in a room full of people that is active and bubbling with excitement. Others prefer the dark seclusion of a safe closet. It depends on how a cat was raised and how secure they feel in given situations. A young kitten raised in a noisy family becomes familiar with the energy therein, and can feel calm when the tempo is raucous. If the kitten was threatened or taunted, however, she may grow up tentative, mistrusting, and wary, and have a tendency to hide when company is around. She may simply not trust people at all. If a cat grew up in the wild, her feral instincts may mean she is cautious all her life, or she may surprise you and lower her barriers, becoming domesticated in the process.

Is his early experience with people what allows a cat to become socialized? In two studies in the 1980s, Temple University researcher Eileen Karsh and colleagues found that cats handled by people at between 3 and 7 weeks of age tend to be friendly and social. But if the felines make it past 7 weeks without human touch, they'll be the skittish kitties that hide under the bed when guests come over.[1] Cats are both predictable and unpredictable. Hence, they have earned the description 'mysterious.'

It is important to determine the personality of a cat as well as learn his language. This knowledge will help you speak 'Cat' faster than you can say 'Get down from there!'

[1] http://www.livescience.com/40708-secrets-to-cat-personality.html (Ghose 2013)

The One Minute Cat Manager advises you to take one minute to check in with your cat. What's going on for them? How can you tell if your cat is happy, sad or angry? There are many things you can look for and spot in under a minute. These are the first steps to master in becoming a One Minute Cat Manager, and learning the language of 'Cat.' It takes just a minute to –

LOOK into her eyes (from a distance)
Cats often feel threatened by staring, and can interpret it as a sign of aggression.

Are the pupils constricted? If so, she can be feeling aggressive, or even contented. Watch for a moment to determine which it is.

Are the pupils dilated or enlarged? She may be feeling nervous if they are really enlarged, or submissive if only slightly dilated. If her pupils are fully dilated she may be on the defensive, or simply in a playful mood and ready to have some fun. Watch for a moment before engaging in playtime.

LISTEN to his voice
Felines have a unique language of their own, and use this to communicate with each other and with people. What they vocalize lets you know how they are feeling.

If you hear growling or hissing, this is a cat who is angry, annoyed, fearful or feeling threatened, and who could act aggressively. Keep your distance.

Purring can be a sign of contentment, or a sign of anxiety (and a sign that a cat is comforting herself, and is similar to a child sucking her thumb). It may be demonstrating happiness and contentment, or it may be letting you know she is unhappy or sick. Observation will tell you what the cause of the purr really is.

If you hear a howl or a yowl a few things might be going on. The cat might be in mating season (if they are unneutered or unspayed), or in distress (trapped in a room or closet). He could be suffering from cognitive decline, as sometimes happens in elderly cats. Take a moment to find out what's really going on. (Don't scold her as she is simply speaking in Cat.)

A 'meow' is an all-purpose sound, according to the American Human Society. "Your cat may be saying 'meow' as a greeting ('Hey, how ya doin'?'), a command ('I want up, I want down; more food now'), an objection ('Touch me at your own risk'), or an announcement ('Here's your mouse'). Some people have experienced their cats walking around the house meowing to themselves.[2]

A mother cat (queen) instructs her kittens to follow her with chirps

[2] http://www.humanesociety.org/animals/cats/tips/cat_communication.html

and trills. If the chirp or trill is aimed in your direction, it probably means your cat wants you to follow her, usually to the food bowl. If you have more than one cat, you'll often hear them converse with each other in this way.

Chattering, chittering or twittering are the noises your cat makes when he's watching birds or squirrels. Some experts think that this is an exaggeration of the sound a cat makes when he grabs prey by the neck[3] (the 'killing bite'). A cat has a natural instinct to hunt, so be prepared for this.

Certain breeds are more prone to vocalizing than others. Other cat sounds you may hear are: moroow, mraaw, mauw, meep, marep, maow, maaaah, muh, muurp, aaaah, raao, yeoy, yo-oy, waaaarah … the list goes on.

If you enjoy peace and quiet be aware that the most vocal breeds are: Oriental, Tonkinese, Singapura, Maine Coon, Burmese, Japanese Bobtail, Siamese, Siberian, Turkish Angora, American Bobtail, Ocicat, Balinese and Sphynx.[4]

OBSERVE her behavior and watch her body language. Felines communicate with their entire body: eyes, tone of voice, tail movement, and the way they position their ears.

Ears

If his ears are forward it means that your cat is alert, interested, or happy. If the ears are facing backwards, sideways, or flat like an airplane it means he is irritable, angry or frightened.

If the ears are swiveling like radar it means he is attentive; listening to every tiny sound in their surroundings.

Tail

If the tail is erect and the fur is flat, this telegraphs that she is alert, inquisitive, or happy. If the fur is standing on end this indicates she is angry or frightened. If the tail is held very low or tucked between the legs, your cat is communicating that she feels insecure or anxious.

If the tail thrashes back and forth, the signal is one of agitation: the faster the tail, the angrier the cat.

If her tail is straight up and quivering, this tells you that your cat is excited and really happy. (If your cat isn't neutered or spayed, this also might indicate he is getting ready to spray.)

Body positions

If the cat's back is arched and his fur is standing on end, this means that your cat is frightened or angry. If his back is arched and his fur is flat your cat is welcoming your touch.

[3] ibid

[4] http://messybeast.com/cat_talk.htm

If you see your cat lying on his back and purring, that's one very relaxed kitty ... if, however, he is lying on his back and growling, know that kitty is upset and ready to strike!

Other behaviors

Rubbing

When your cat rubs his chin, head or body against you, he is marking his territory. You'll notice that he may also rub the chair, the door, his toys ... everything in sight, in fact. This is how cats claim their territory and possessions: marking you with his scent is a sign of affection and possession.

Kneading

Also called 'making biscuits' because a cat works her paws as if kneading bread dough, this is a trait from kittenhood when a nursing kitten massaged her mother's teats to make milk flow. Your cat expresses happiness when she does this.

The Flehmen Response

Have you notice a cat lifting his head, mouth slightly open, curling his lips and squinting his eyes when sniffing something? A cat's sense of smell is so important that they actually have an extra olfactory organ that very few other creatures have: the Jacobson's organ (vomeronasal organ), located on the roof of the mouth behind the front teeth, and is connected to the nasal cavity. Cats open their mouth and inhale so that the scent molecules flow over the Jacobson's organ, which intensifies the odor and provides more information about the object they're sniffing.

Personality

Take a minute to figure out your cat's personality type, so that you'll have an idea what sort of behavior might be breed-specific.

Have a go at figuring it out according to the characteristics listed below –

Is your cat ...

- Dominant
- Spontaneous
- Skittish
- Outgoing
- A hunter
- Friendly
- Affectionate
- Curious
- Solitary

Assigning a percentage value to each quality will lead you to a clearer picture of kitty's personality type. You can even find tests on the Internet that will tell you what kind of a cat you have.

One of the best ways to determine what personality your cat has, or his tendencies to be outgoing and gregarious, or shy and withdrawn, is to check out the guide on the Cat Fancier's Association Website http://cfa.org/Owners/FindingAKitten/BreedPersonalityChart.aspx.

This is a thorough guide to cats and their personality type. If you have a mixed breed, you can take the suggestions given and include a few others, because multiple personality types will be mixed together.

You can also take the cat personality test on the web http://www.knowyourcat.info/quiz/catcharacterquiz.htm.

Knowing your cat's personality type will benefit you and your cat, as you will be able to recognise breed-specific behavior when it occurs, and understand why your cat is doing it. The knowledge also assists when selecting appropriate amusement for her, and explains their habits and tendencies, all of which makes for a better relationship for you both.

Is your cat feral or a stray?

A feral cat is an animal who has either never had any contact with people or whose contact with them has diminished over time. He is fearful of people and survives on his own outside. Because of her condition and life patterns, a feral cat is not likely to ever become a lap cat, or enjoy living indoors. However, kittens born to feral cats can be socialized at an early age and adopted into homes.

A stray is a cat who has been socialized to people at some point in her life, but has left or lost her domestic home, as well as most human contact and dependence. Over time, a stray cat can become feral as her contact with people dwindles. However, a stray cat can also become a pet once again under the right circumstances. Stray cats who are re-introduced to a home after living outside may require a period of time to re-acclimatise, because they may be frightened and people-wary after spending time outside away from people.[5] It only takes a minute or two to assess what your cat is trying to communicate. Once you learn the basic tenets, you'll speak 'Cat' for the rest of your life, and create a rapport with kitty that will bring both of you enormous joy and years of happiness.

One Minute Cat Manager summary

- Look, listen, and carefully observe your cat
- Understand that cats have a different way of communicating
- Observe her body language
- Learn to speak 'Cat'
- Try some cat language and see how well you can communicate

[5] https://www.alleycat.org/resources/feral-and-stray-cats-an-important-difference/ (Alley Cat Allies nd)

3 Understanding cat behavior

The Singer

When a cat wants to sing a song to you, your job is to listen for one, full minute. The concert may go on longer, but for that single minute your focus must be on kitty; praising him or her for their glorious vocalizations. Your job is only to appreciate and listen, no matter how painful or delightful it may be.

At our house there lives a gentleman cat named Percival, or 'Peevie' for short. During the night, usually between two and four in the morning, Peevie enchants the entire house with a free concert. We call him Peevoratti during those pre-dawn musical adventures. The concert is usually a celebration of his most recently-nailed capture. It could be the fuzzy purple mouse freed from under the couch; it could be the travel pack of Kleenex stolen from an open purse, or it could be a felt marker left over from making signs for the church bake sale. He doesn't discriminate. If it looks fun, Peevie nails it.

When Peevie locates his prey he totes it to a new location, and the joy of the victory march causes a song to burst forth from deep within his warrior soul. He sings his little heart out in those moments of triumph ... and who are we to rain on his parade? Mere mortals! Thus, we have learned to be complimentary, congratulatory, appreciatory ... and then tactfully suggest a nice rest might be his next best option. But only when he has completed the aria and accepted his applause. We always pray he keeps his concerts to the one-minute format.

Often, your cat will need an audience, and will seek praise for bringing you a gift, showing you a trick, or demonstrating some death-defying agility. Other times, kitty will *become* your best audience. If you are prone to

The Entertainers

humming a tune or belting out a ballad, your cat-critic may run like the wind to the farthest corner, or she may sit like a little princess and listen to your every sound (in some instances, she may join in). This trait is particularly helpful when you are rehearsing your speech, practicing your closing remarks or developing your keynote address. Kitty will listen and let you know with an unabashed yawn if you are boring her. This feline feedback will teach you how to remain present, connected, inspiring and authentic. Your presentation will be much improved with cat input.

Presenting you with a prize

Cats *love* to bring you presents. Your job is to accept them with graciousness and gratitude.

"Cats learn through experience and traditionally they are raised by their mothers. This involves teaching the kittens how to look after themselves, including how to catch prey. Given that female cats are most likely to bring back animal presents, the most likely explanation for this behavior is that they are trying to teach you the hunting skills that you clearly lack."[6]

To reject their gift or to criticize them is confusing to the cat, who is simply attempting to teach you what is appropriate for survival in the cat world. Their instincts include hunting and gathering the fuzzy purple mouse you gave them for Christmas, or the stuffed blue felt bird they got for Hanukkah.

The One Minute Cat Manager encourages you to support your cat's generosity, clean up the mess and not scream blue murder when you find a small rodent carcass on your doorstep. Remember: it was brought to you with love and the very best intentions. Kitty thought it was a really good idea. Be generous and praise her.

Attention-seekers

Attention-seekers persist in certain behaviors because they achieve the desired results. If it's attention they crave they may choose annoying or adorable methods to get that attention from you. Cats are quick to figure out that their tactics work because their people reinforce and encourage the behaviors by interacting with them, whether or not they are aware of it.

The next time your kitty gets your attention, take a moment to figure out how they did this. Did they do something adorable, or did they knock something off the table to startle you? You have to use ordinary psychology to change the behavior if it's not something you want repeated.

Cats test out things; this is part of their learning process. If the cat did something negative to get your attention, then you must ignore the action and give them no attention whatsoever. (No scolding; no swearing.) If they did something pleasing, reward, pet and praise them for that action. You may have to remove enticing objects from their reach while you are retraining them, but cats are quick learners, and a few attention-seeking cat-acts that garner no response will soon ensure that kitty gets the message.

[6]http://www.iflscience.com/plants-and-animals/why-do-cats-bring-home-dead-animals/ (Janinski n.d.)

Reactions such as yelling, threatening and chasing can backfire and strengthen behaviors instead of halting them. Positive or negative reactions have one thing in common: both involve human parents responding to the cat's behavior by interacting with them. From the kitty's point of view, when they do specific behaviors, their people pay attention to them. Treats, affection, grooming and encouraging words are examples of good behavior reinforcers that work for individual felines. It's up to you to reward the behavior you like and ignore the behavior you don't like. Nobody gets hurt this way and relationships will thrive.

The Acrobat

Many cats enjoy climbing onto high places and leaping to a lower level. The word 'catapult' was coined to illustrate this activity. Kitty may think your house was designed by the Wallendas, and that flying through the air with the greatest of ease should be a daily ritual. You may feel differently, however ...

Cats are 24 million years old. They descended from Proailurus, the first true cat. Our current domesticated cats trace from the Middle Eastern wildcat, Felis sylvestris (which literally means 'cat of the woods') 12,000 years ago.

The original cats came from the forest and the woods. They perched in trees to spot their prey (dinner), and used the foliage to conceal their catch from airborne predators such as hawks and eagles.

House cats enjoy climbing into high places because it gives them a view of their environment. They enjoy shelves (why not leave a high-mounted one free of books and knick-knacks for kitty to nap on?), cat trees (especially when they are next to a secured window), tall cabinets and ladders. Outdoor kitties enjoy climbing in trees and onto poles for the same panoramic view.

The other benefit of being on high is warmth: since heat rises, your cat may be seeking the warmest place in the house.

The One Minute Cat Manager suggests you create some spaces and places where your cat can climb to safety and spend time alone or with a friend enjoying the view from the top. You can go so far as to install an upholstered shelf for comfy napping and viewing.

Just remember to keep your valuable display items and travel souvenirs on a shelf that does not beckon to kitty.

Scratchers and shredders

Who's more important, kitty or your sofa? Actually, you don't have to make a choice: you can have both. The first thing this book recommends is that we recognize cats have an instinctive need to scratch and stretch their claws. It's a primary cleaning and marking need that all cats have. There are ways to channel this instinct and prevent furniture damage.

Cats do not sit up all hours at night trying to figure out ways to get people to shriek "No!" in five different languages.

Outdoor cats use trees, stumps, grass, poles, and whatever they can get their claws into. Indoor cats are more restricted, and can only access what is at hand (or at paw, as the case may be). Hence, the sofa or your dining room chairs become easy targets if there's nothing else. The clawing and scratching activity relaxes the cat, relieves stress, stretches front and back leg muscles, aligns the spine and back, and serves the purpose of marking territory. Scratching really is not a pre-meditated act of war on your furniture.

There is a simple fix. Cat Trees are available to satisfy their clawing needs. Corrugated cardboard scratchers are available, as are jute scratching panels to hang from door knobs, and a whole host of other creative DIY ways to create a place for kitty to satisfy her needs and stretch her claws.

Please DO NOT even think about declawing your cat, in which the last bone of each toe is amputated – similar to amputating the last finger joint. This is barbaric and painful – and illegal in some countries where it is considered inhumane.[7]

Cat kisses

Non-verbal communication is the way of the Cat. They use their eyes to communicate trust and affection. The equivalent of a cat kiss consists of fixing your gaze directly on your cat's eyes and, in a long, slow movement, blink whilst looking at him. If your cat reciprocates, he's showing you that he trusts and loves you. You can double the effect by sending even more affection his way by verbalizing, 'I love you' while you extend your cat kiss.

Birthdays

Usually, a vet can more-or-less determine the age of a cat or kitten by examining her and giving you a kinda-sorta estimate. If you don't know the exact birthday of your cat, you can approximate it. *The One Minute Cat Manager* encourages you to celebrate your kitty's birthday by throwing an elaborate party, or having a small, thoughtful and intimate celebration of your feline's natal day.

We've seen everything from engraved invitations, elaborate decorations, catered cat meals, and hired limo drivers to a small can of tuna opened in the kitchen in honor of the birthday cat. It doesn't matter what you do; it's always a lovely idea and a considerate gesture to remind kitty

[7]http://www.humanesociety.org/animals/cats/tips/declawing.html

you are celebrating the day they were born. One small birthday song (and can of tuna) goes a long way in a cat's world. Kitty will appreciate your kindness.

Holidays

In some multiple cat households, the cats send Valentine cards to each other. In others, the day passes without acknowledgement of the day. Some parents dress their cats like leprechauns for St Patrick's Day, whilst others save the green beer for themselves.

National Cat Day is October 29 every year, so you may want to mark your calendar with this date.

Christmas and Hanukkah are irresistible opportunities to wrap up a few presents for kitty, and maybe even hang a stocking with his name on it by the fireplace.

In general, cats can find many playful opportunities with, and in, the Christmas tree and low-hanging decorations. This is the season to be particularly aware of the dangers that lights, globes, glass ornaments, and wire ornament hangers can pose for the health and safety of your cat.

Candles, flames and fireplaces can entice a cat: make sure you cat-proof your holiday décor so no one has to spend any time at the emergency hospital.

Vacations

If you take a vacation or two each year, kitty must be cared for, too. Is that something your neighbor can do, or does your feline need to be cared for in a cat hotel or boarding place? If you are going to board your cat, make sure her shots are up to date at least two weeks before you leave so that any reaction to the shots can be monitored. It only takes a minute to book the appointment.

Your to-do list of trip preparations should include the health and

Take me with you!

welfare of kitty. Will you need to supply the food? Will your cat get enough exercise? Are there times when your cat can be played with? If you are having a neighbor or a local teenager come by, have you left adequate instructions, food and supplies (such as litter) for your absence? Often, young people don't have experience of caring for an animal, and this could result in dirty litter, unwashed dishes, bugs, or empty water dishes. Be sure you rehearse the feeding ritual with your designated feeder, so kitty doesn't suffer from the errors and omissions of a novice. You might also arrange for another adult to check in to make sure your cat is being well cared for.

Illness

Your cat is designed to be long-suffering. In the wild if they showed any weakness they could become prey, be left behind, or killed. Hence, kitty is going to do everything in his power to hide any symptoms of illness – which sometimes makes it hard for a cat parent to notice when he is sick. Keep regular wellness appointments with your vet, and also spend a moment, at least once a month, observing kitty in the following areas –

Kitty under the weather

THE OBVIOUS SIGNS

Bad breath: This is a sign that your cat may be having internal issues
Vomiting: Beyond hair balls this signifies a greater problem
Dilated eyes: Tells you something needs attention
Shallow breathing: This requires immediate attention – and may even be an emergency
Blood in the urine or stools: Several organs could be affected and causing pain
Eliminating outside the box: If the cat box is clean and scooped out, your cat may be telling you something is wrong when he uses the floor instead of the box for his bathroom duties. He could be indicating internal organ problems, or even an internal blockage. Get help. By checking for these signs regularly you can avoid more serious conditions that will develop if let unattended

THE NOT-SO-OBVIOUS SIGNS

Hiding: If your cat begins to hide a lot, she is trying to avoid becoming prey. Get her to the vet
Grooming: If you see bare patches of fur, a greasy coat, lots of matts, something is wrong

Weight change: Is kitty not eating or drinking as usual? This is a sign that medical attention is needed
Activity change: If kitty stops being active and sleeps more than usual, she needs to be checked out
Behavior: If your cat goes from being wonderfully social to withdrawn and reclusive, he's hiding a symptom
Voice: If your feline's voice changes (becomes weaker or louder), something abnormal is going on
Sleep: If your kitty's sleep pattern changes and she exhibits different behavior at night or during the day, she's asking for help

The pay-off

Cats are non-stop entertainment, comforting sources of affection, and intrinsically natural comics. They can be great company, or even alarm notifiers. They are loyal and protective, and some feline authorities even believe that cats regard us as their offspring, and feel an instinctual obligation to train us in the ways of survival, and the Way of the Cat. According to author John Bradshaw in his book Cat Sense, cats treat humans like other cats.[8]

The One Minute Cat Manager suggests you accept that you are a cat in their eyes and treat your cat like a person, which should make for a very happy and lasting relationship ... with the cat, obviously, in charge ...

The One Minute Cat Manager summary

- Make sure your cat has special areas to go and structures to climb
- Be sure your cat feels part of the family, and include him in holidays and celebrations
- React with calm and appreciation to the kindness kitty shows you
- Put kitty's needs on your to-do list when making travel plans
- Learn how to 'kiss' like a cat
- Observe the obvious and not-so-obvious well-being signs from your cat

[8]http://new.www.huffingtonpost.com/2014/01/15/john-bradshaw-cat-sense_n_4603722.html (Mosbergen 2014)

4 How to change your cat's behavior

Merlin Rose was a petite runt-of-the-litter kitten. We were told by the vet that 'she' was a 'he,' and, disguised as a him, we were duped into adopting her.

Merlin Rose began life in a trailer park in Arizona. A pregnant mother cat was rescued by a compassionate animal-lover, and taken to the local vet, and a few days later four kittens were born. The vet team looked after them until they were six weeks old and ready for adoption. Our friend living in Arizona thought this small, grey cat was a perfect substitute for our beloved Jabez, who we lost the year before: 'he' was the same color, and seemed to have the same personality, she said. She encouraged us to adopt this replacement, promising she would get him transported to LA for us.

Merlin arrived as a teeny-tiny little waif of a thing. We constructed a temporary, circular, open-top, 3-feet high enclosure for him in the living room so that he could gradually integrate with the other cats. The sign on the enclosure read 'Merlinville: Population 1.'

Merlin was like a mini pewter-colored rocket. No sooner had we put him in the spacious enclosure than he scaled the sides and bolted. Nothing would keep him inside except for the spare door we kept in the garage, plus the extra leaf from the dining room table that we strategically placed on top of the wire enclosure. We wanted to keep the tiny tot away from the others for integration purposes, and because he arrived with the sniffles.

He was so little; looked so pathetic, and had a cold. We made a sling for him and one of us 'wore' him in the sling all day for warmth, and to relieve the separation anxiety he may have experienced leaving home. We knew he'd been low man on the totem pole in his litter because he was the runt and seriously puny.

Merlinville: Population 1

Merlin integrated pretty well into the household. He was always trying to make up for his size by showing off his might with a grandiose attitude.

Six months into the adoption we realized that Merlin was not a he but a she. The local vet confirmed it and everything changed. Merlin became Merlynn and overnight we had four girl cats and one boy.

Merlynn had learned early on to be demanding, vocal, and kind of a pain in the neck. She cuddled at night with her people, but never formed one good cat friendship: even now, it's mutual toleration at best.

Last year, she decided that she wanted to be fed more often than twice a day, and at noontime began caterwauling and complaining until we caved in and gave her a couple bites of cat food. We would remind her to use her 'pretty' voice when she wanted something instead of her screechy 'Fran Drescher' voice.

After a month of this, we knew it had to stop because it was disruptive and annoying, and unnecessary to boot. We didn't want to deny her food if she was *really* starving, but, knowing Merlynn, this was yet another dramatic plea for attention. After a vet visit to make sure nothing was wrong, a lightbulb went off and *The One Minute Cat Manager* kicked in.

We decided to change her experience. When she began to vocalize her demands we simply picked her up and cuddled her instead of feeding her. We rocked her like a baby, petted her silky head, and told her how special she was. This approach must have taken her back to her early days and did the trick. It was nothing short of a miracle. The whining ceased, and peace was restored throughout the kingdom.

It is possible to change behavior with just one minute of focused distraction and comforting care.

Many cat habits can be changed from destructive to constructive with positive reinforcement and a minor investment of time. The key word is consistency. Be sure you only reward the desired behavior and ignore the undesired pleas for attention, or drama.

The theory behind training a cat begins with the familiar adage, "You can catch more flies with honey than vinegar." Never punish; only reward.

Methods

If you want your cat to repeat a behavior, reward him for doing it.

If you want your cat to stop a behavior, ignore him when he does it; give him zero attention, and then reward him when he does something you do like.

To get your cat to come when you call, call his name. If he comes to you reward him: do this several times. Change locations. Call again and reward him when he comes. If he doesn't come to you right away, have patience and wait for himto approach you. Speak softly and lovingly, and indicate that you have a treat for him (*don't* disappoint).

If you want kitty to scratch his scratching post instead of your sofa, reward him when he scratches the post, ignore him when he scratches the sofa, and put double-sided sticky tape on the undesirable places where he scratches to make it unpleasant to scratch there.

Rewards speak louder than words

Food treats are usually the best motivators. Cats quickly link behavior with a treat reward.

Cats have short attention spans so reward right away, within seconds, to anchor the reward to the behavior. All family members should be on the same page and repeat the same action.

Be consistent. Give the same kind of reward every time your cat exhibits the behavior you want to see. Every time she repeats the behavior you like, give her a treat. (Yes, you may have to carry treats around with you in your pocket, but it won't go on forever, and it's a small price to pay for well-behaved kitty.).

Cats learn by trial and error and they are smart. After they perform a behavior, they evaluate the results. If the behavior results in getting them what they want, they will repeat the behavior in the future.

Training time is best right before meals when your cat is hungry, making the treats the most valuable. Look for low-fat cat treats so kitty doesn't end up the size of a blimp.

You can begin to phase out the food treats and replace the reward with praise and petting. This change will require more time to accomplish. Start by using a treat, then praise or a pet, then the next time the good behavior occurs supply a treat. Vary this ratio 2-1 then 3-1, 4-1, etc, until you have phased out the food treats altogether. Give an occasional treat, though, as kitty deserves rewards!

Sometimes a treat following a procedure such as teeth-cleaning, brushing or nail trimming, will help ease the stress of the event. We use a variety of treats, and a hair ball preventative in a squeeze tube (tuna-flavored) as a reward.

Four things you should never do (aversive conditioning) –

- Never force your cat to enact the behavior you want. Don't pick him up and take him to the scratching post, litter box, or food bowl. He is a cat! Behavioral training is only effective when it's his idea and he performs the action. When he receives a reward for a new behavior, he begins to associate the action with the reward
- Never punish your cat. He will not be able to associate the action (broken vase, chewed shoe) with the punishment. Cats retain a memory for a few seconds only, and punishent will result in a broken bond between you and your cat, who will now be fearful and confused
- Never yell at or scold your cat. This may be tempting after the discovery that your favorite mug has been shattered, but your outburst will only terrify your cat and make him frightened and anxious
- Never spray your cat with water. If you want your cat to get down off the counter, don't hit him, spray him or yell at him. This is aversive conditioning

and only results in your cat fearing you and associating you with harm to himself

The joy a cat can bring you far outweighs any problems. In my house we say, "Oh well," clean up the mess, and move on. Once the shock has passed, humor steps in, and everyone can find a chard of kindness. I remember the time Jabez scooted the antique French crystal champagne glass over the edge of the table, and was so pleased when he heard it shatter in pieces on the floor. He thought he had accomplished a great feat. I remember standing there in complete shock, just looking at him. As angry as I felt in that moment, when I saw the look of pride and accomplishment on his face, as if he was saying, 'Look what I did just for you,' I couldn't help but say, "Oh well."

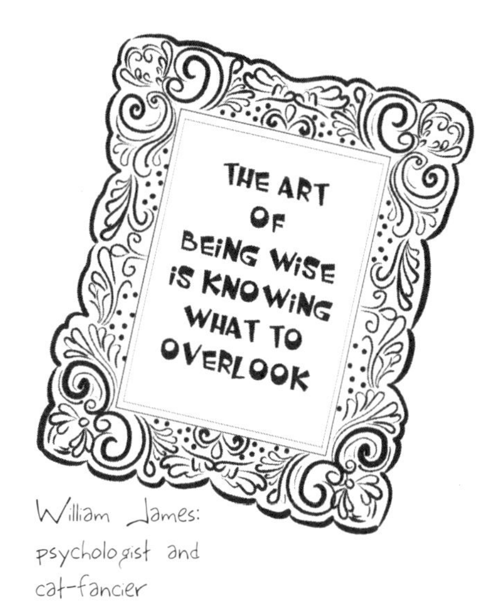

William James: psychologist and cat-fancier

In spite of appearances your cat doesn't think like you do. "What makes a cat appear untrainable is the fact that it will perform what it was trained to do on the basis of whether or not it wants to do it. Because the cat is not a pack animal, there is no inherent need or desire for the cat to comply with anyone's wishes but its own. We humans have a difficult time accepting this because we relate as pack animals."[9]

The One Minute Cat Manger reminds you that you can modify a cat's negative behavior if you are willing to put in a few minutes of your time. Withdraw your attention while the undesirable behavior is occurring, and give attention – followed by a treat – only when the desirable behavior occurs. Always take the high road with your cat.

The One Minute Cat Manager summary

- Cats respond very well to positive reinforcement
- Train a cat to exhibit desirable behavior by using treats as a reward
- They key is consistency
- Never yell at, spray with water or punish your cat for undesirable behavior as your cat will learn to fear you
- Only use positive conditioning as a means to modify behavior

[9] http://www.perfectpaws.com/help2.html#.WJ5uw_krJaQ

5 Funny feline foibles

We expect cats to meow, purr, rub, play, climb and chase. But do we expect our cats to have a love affair with a rug, crave cantaloupe, or steal Kleenex packs out of your purse and hide them in a secret nest?

Many of the joys of living with cats are the precious things they do. Besides secretly watching cat videos on the web, take a minute to start up a conversation with your cat-loving friends, and hear what they have to say about their kitty and the odd things they do. Cat stories can entertain you for days.

Percival loves rugs. He can be found not just rolling around on one, but slipping his paw under the corner and cuddling it like a sweetheart. He purrs, sings to the rug and licks it. Our first response was, "Yuck, Percy, that's a rug!" but then I realized we had hurt his feelings. We still have no idea why he loves rugs so much, but, on any given day, you can find him caressing the bathroom rug, the living room rug, or the purple rug in the den. It's just 'his thing' so we leave it alone and try our best not to judge.

Jazzmine steals Kleenex packets out of purses in the wee hours of the morning. She gathers them like kittens, and takes them to a nest she has created. Percival finds them, takes them out of the nest, then meows at the top of his lungs in a victory cry for having rescued the lost souls. (Who needs Shakespeare when you have this nightly drama going on?)

Dot loved cantaloupe, waffles and pancakes. No matter where he was in the house, as soon as he sensed or sniffed any of the aforementioned, he was down to the kitchen like a shot. Cantaloupe was his ambrosia, and pancakes and waffles ... I mean, who isn't crazy about them?

Cats can be aficionados of your feet and slippers. They respond to the molecules of scent, and can associate you with the aroma. If they snuggle

Cat burglar

your slippers, it's because of their feelings of affection and familiarity.

Lucy loved to sing. Her favorite song was: *Sing, Sing a Song* by The Carpenters, and she would meow (loudly!) at the end of each line. Many a house guest was treated to after dinner entertainment featuring this well-known seventies song as performed (warbled) by her mom, with Lucy punctuating each stanza.

Mandy is visited by a coterie of imaginary friends. She chats with them, naps with them, and is frequently seen standing in the window with them describing the view. We have long since stopped interrupting her when she is entertaining her friends., and she's very appreciative when we invite them to stay for dinner.

Harmony was the artist-in-residence and interior decorator, maybe left over from a past life. Often, she helped herself to any houseplant that was gracing a table and relocated it to a place she thought was better. Many a day, after work, we would find an empty planter or pot, with the plant, roots and all, hauled to a new location. The planting compost on her paws always gave her away ...

Cardboard boxes are sought-after treasures for cats. Jazzmine loves sitting in

My friends are real to me

them; Merlynn is transformed by simply being around cardboard, and Percy believes they are hiding places for cat treats. Jazzmine and Merlynn also adore baskets piled high with clothes fresh from the dryer.

As a kitten, Jazzmine liked to remove her dry food from the bowl and place it on the floor. We don't know if she was separating by color, alphabetizing or categorizing them, but at each meal she systematically removed 10-12 pieces of kibble before eating them. It was her ritual. She was extremely tidy and never left a piece on the floor after she was done. We never figured it out.

Socrates has a penchant for sinks, and you'll find him most days (when not asleep in the dog's bed or in the middle of my desk,) resting and posing in a sink. The prettier the sink, the better, as far as Socrates is concerned.

Socrates and Merlynn have an unnatural affinity for plastic – plastic bags, plastic containers, plastic substances of all shapes and sizes: all get a good licking.

Elegance is my middle name

Lucy (and I almost hate to tell this story) loved to lick things that weren't hers. One year it was the sheets of stamps I had purchased for my Christmas cards. I came home from work to find that all of the sheets had been licked dry of any glue they had had. (This was in the days prior to self-adhesive stamps.) That same year, I had been given a beautiful 'chocolate' house (think gingerbread house only made from chocolate) from a co-worker on a TV food show, and Lucy licked off half the roof in the middle of the night. Naturally, I was terrified that the chocolate would harm her since chocolate is one of the big no-nos for cats, but my vet told me to observe her, watch for vomiting and diarrhea, and call back if symptoms showed up. Fortunately for us, nothing did but, from then on, I covered up and stored away anything chocolate in the house since chocolate is lethal for (most) cats. But then Lucy was not 'most cats.'

All the cats in our house have been given their own unique song. Usually, it's a well-known tune with personalized lyrics. Mandy is somewhat roundish, all white with peach dustings; part Angora, part Siamese, and part Himalayan. As a result, her song is, *Mandy the Snowball* (tune: *Frosty the*

Snowman). Percival's song is based on the theme tune for HR Pufnstuf (a US children's television series); Merlynn's song is sung to the tune of *Yankee Doodle,* and Jazzmine, bless her heart, has a song that is sung to the tune of *Ramblin' Rose*. Socrates, the Ragdoll, is serenaded by *The Most Beautiful Boy in the World*, because, well, he is!

We like to sing in our household, so if you're passing by someday you might hear a familiar tune with lyrics re-written especially for a cat. Sometimes the cats have been known to sing along with us.

Each cat has unique characteristics and a personality all their own. Yours will entertain you with their special talents and peculiarities. *The One Minute Cat Manager* encourages you to enjoy each antic, and join in if you can to show kitty you know they are fabulous and indeed, very, very special. You are allowed to act daft when you share your home with a cat.

The One Minute Cat Manager summary

- "The smallest feline is a masterpiece" – Leonardo da Vinci

6 Time-out for kitty: Spa Day

Many of us are under the impression that a cat is self-sufficient, and we don't have to worry about grooming her as she takes care of this herself. We can easily get caught up in the received wisdom that all we have to do is open a can of cat food and empty the litter box – here, kitty, kitty – done. *The One Minute Cat Manager* approach also applies to grooming habits, and this is a minute that really counts.

A well-groomed cat is a healthy cat. Cat grooming does not have to be extensive, and the rewards are incredible. Every other day, a one-minute combing can tell you as much about the future of a cat as a reading of tea leaves. Cats do like to groom themselves, it's true, but some are more fastidious than others. One tip-off that your cat is ill might be when they stop grooming themselves or begin to look greasy. Matted, greasy hair on cats will not only have an unpleasant smell, but, in some cases, can be linked to dental problems, bladder infections, diabetes, or other medical conditions

By giving your cat a one-minute groom, you'll also be able to spot other problems with your cat's skin such as insect bites, allergies, a flea infestation or acute dermatitis. Always comb gently; do not pull kitty's hair and look for any area that indicates attention is required. If your cat over-grooms himself this may be a sign of deeper issues such as food or pollen allergies, dermatitis, or even neurological disorders. Check out this website: https://www.vet.cornell.edu/departments-centers-and-institutes/cornell-feline-health-center/health-information/feline-health-topics/cats-lick-too-much

You will need to have a few very affordable supplies on hand –

- A two-sided stainless-steel pet/cat comb

One side is for thicker mattes and the other is for smoother, untangled coats.

Use this once a day and you will keep your cat's fur tangle-free. It only takes a minute.

- Hairball remedy – usually purchasable in a tube or as a component in a dry cat food mix

There are some natural ways to help your cat get rid of hairballs, which are a by-product of their instinctual grooming habits. Add 1-2 teaspoons of canned pumpkin to your cat's wet food three-four times a week, or add 1 teaspoon of olive oil to your cat's wet food once or twice a week. Both help push those hairballs to the south end and not the vomiting end.

- Kitty shampoo

You will want to have a good natural kitty shampoo on hand. You never know when your cat might end up sticky, greasy, smelly, or covered in something you really wished she wasn't. Never use human shampoo or any other kind of soap on cats because their hair and skin have a different pH level to ours. Perhaps you have seen television coverage of wildlife rescues and cleaning up after an oil spill? They often use a dish-washing liquid called Dawn. Dawn (or Fairy Liquid) has excellent grease and oil-removing abilities, and can also help remove the odor caused by a skunk, should your cat have been sprayed by one. These dish-washing detergents, however, are inappropriate for regular bathing of your cat because they will quickly strip her skin of the natural oils that help to nourish and protect it.

The best shampoo for a cat is one that has been formulated especially with feline pH balance in mind, and is as organic and natural as you can find. Use warm water, never hot, and try to avoid getting water in kitty's ears or on her head. Place some cotton wool in her ears (but don't push this in) to prevent water getting in and proceed with care.

It's a great idea to keep your cat's claws trimmed just in case there is an emergency need for a bath, and so avoid being scratched (see #4).

- Kitty nail clippers

The final thing you will need is a good set of cat nail clippers. Don't just go into your bureau and pull out the handy clippers you use for your nails; these can pinch and even shred the nail if they are dull. Invest in an inexpensive set of cat nail clippers and you'll be able to trim kitty's nails in a short session.

Take a minute to learn where the 'quick' is on a cat's nail, so you don't inadvertently chop into their blood supply and end up with a hurt, bleeding and unhappy cat ... and/or a shredded arm.

Mandy Rose thinks nail trimming is a real treat. She loves playing 'spa' and has no trouble extending a paw for her weekly manicure.

All of these things (except for the bath) take just a minute or less, and can pay off in the long run with

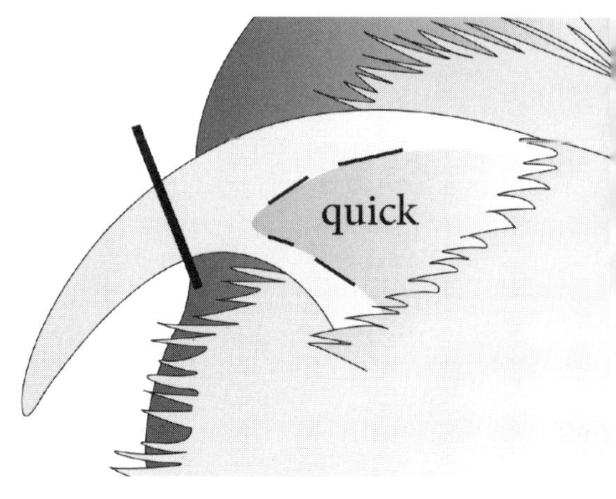

Clip the nail; not the quick

Mani-Pedi, purrlease

kitty's lovely appearance, well-being and bonding time with you.

Please have a look at some of the great research online that can give you a step-by-step picture of how to safely and effectively groom your cat; especially if you have a longer-haired variety, as they are prone to mattes and tangles.

At our house the grooming sessions have become lovingly competitive. The boys push the girls out of line in order to get more combing. I have not yet been successful at teaching them manners, but it is entertaining to watch how they will do anything to be first in line for the combing. Just this morning Jazzmine pushed closed the bathroom door as I was combing her. She wanted to put a stop to intruders and line jumpers once and for all!

How to groom a cat –
https://www.wikihow.com/Groom-a-Cat
https://www.vetbabble.com/cats/grooming-cats/grooming-tips/
https://www.aspca.org/pet-care/cat-care/cat-grooming-tips

The One Minute Cat Manager summary

- Grooming a cat is as essential as grooming yourself
- Create a combing ritual every other day to keep a step ahead of illness and issues
- Bathing a cat does not have to be a life-threatening ordeal
- Some kitties actually enjoy having their nails trimmed: yes, really!
- It takes only a few minutes to learn how to safely groom your precious cat

7 The Zen of cat poo

When you call your vet for an appointment for your ailing cat, they usually ask you for a stool sample if your cat is suffering from gastrointestinal issues like diarrhea, worms, constipation or food allergies. They make matters even worse by asking you to provide the sample free of cat litter. What?! How am I supposed to do that? Have my cat poop on a plate? Have him poop in my hand? What are you asking me to do? Gross!

Of course, the vet wants a perfect, pristine sample, but here's the good news: they'll take whatever you can give them.

If your vet requires a stool sample from your cat –

- Throw out the old litter and put in fresh with a new liner to make it pleasant. Not only will this attract the cat to do his or her business, it will also be uncontaminated for the sample
- If you have multiple cats, temporarily (for half a day and especially after meals) isolate the cat in question and get a stool sample from a separate litter tray. (If you can't isolate your cat or don't have a separate litter tray, one hilarious suggestion I found on the Internet was to feed all the other cats sweetcorn. You'll be able to tell the difference between your ailing cat and the others by whose poop doesn't contain corn. Sounds effective to me)
- If you have an outdoor cat, you'll need to follow him after he's eaten and locate the feces in the area he or she chooses. This turns you into a Cat Poop Detective (you could dress for the part with a long coat and Deerstalker), or maybe contain your cat in a confined area with a litter box for an hour or so after meals until he's provided the sample

The vet usually only requires a sample the size of two sugar cubes, so

you won't need to take in large quantities.

"Murphy's law states that when a stool sample is required, a cat will not poop the morning of a doctor's appointment, except, of course, the one time the cat poops ten minutes prior, in which case it states you will forget to grab it."[10]

Items you'll want to have on hand –

- Rubber gloves for you (or a plastic bag)
- A plastic disposable spoon for collection
- A container such as a clean pill bottle, a small snack bag (double-bag it), a pet waste bag (usually when you walk your dog you take a roll of these with you), or a fecal container provided by your vet. (Be sure to LABEL this for contents ...)
- Store the sample in a cool location until your appointment. Don't let it sit in the sun or be exposed to air for any length of time
- Pin a note on the door to remind yourself to take the sample with you
- Be sure to wash your hands with soap and water after you've collected this sample

If you are having difficulty getting your cat to comply, here's one suggestion from a fellow cat owner –

"You could try chanting 'poop poop poop, move your bowels' over and over again and telepathically suggesting that your cat empty its colon. I've never found this to help, however, although it may entertain you to watch the facial expressions of others in your household as you do it."

The One Minute Cat Manager suggests that you spend one minute twice a week checking your cat's poop. I know it sounds like a task you'd rather not do, but detecting a problem early may improve your cat's health and prolong his life.

What to look for –

"Just like diamonds that are evaluated by the four Cs – color, clarity,

[10]http://pets.stackexchange.com/questions/5829/how-can-i-encourage-a-kitten-to-poop-for-a-stool-sample

carat weight and cut grade – stool samples are evaluated by the following: color, shape, consistency, size, and content."[11]

- The color of the feces should be brown (the exception is if you feed them a diet with colored dry food).
- The shape should be that of a log (unless your pet has an obstruction, hip damage or arthritis)
- The consistency should be like dough so that when it is picked up it doesn't lose its shape
- The size should be relevant to the size of your pet. The stool should be consistent and if it changes in size, something internal may need attention
- The content should be consistent with what the cat eats. (if you spot anything that wiggles (like worms), see undigested particles of food (like rice, carrots, beans), detect mucous or blood (which could indicate inflammation), or anything that shouldn't be there (the list spans toy parts yarn, crayons and so on), it's time to call the vet and make an appointment

Take a minute, inspect the poo, and save your pet from pain, ill-health, or worse.

The One Minute Cat Manager summary

- No need to be intimated by collecting a stool sample. You only need a small amount for the lab analysis
- Store it correctly
- Put a reminder on the door not to forget it
- Use gloves to keep you and the sample sanitary
- Check your cat's poop once a week for good health
- Take her to the vet at the first sign of any irregularities
- Use the guidelines in this chapter

[11] http://pets.stackexchange.com/questions/5829/how-can-i-encourage-a-kitten-to-poop-for-a-stool-sample
http://www.huffingtonpost.com/donna-solomon-dvm/pet-poop_b_2464601.html

8 Packing *for* cats and packing *with* cats

Those of us who have a cat know what packing for a trip is like with a cat around. Curious and always in search of a new place to nap, the open suitcase is a lighthouse with a catnip beacon for them. Pack the jeans; remove the cat. Add some tops; remove the cat, and so goes the drill as you attempt to keep your focus on what to take for which occasion, and how many. I've arrived in many a city with fresh cat hair on my dry-cleaned clothes, and the occasional cat toy tucked in the side pocket. Which is exactly why I always pack a lint remover or two.

Stowaway

Do my cats know I am leaving; is this a ploy to make me feel guilty? Are they out for revenge, or are they sad I am going to leave them and want me to know it?

There is a technical term for what some cats feel when their person leaves them: separation anxiety. This syndrome is more usually associated with dogs, but domesticated cats can experience it, too.

"Cats with separation anxiety may express their dissatisfaction by eliminating outside the litter box, spraying urine on your bed or clothing, throwing up, grooming so compulsively that they develop bald spots, or scratching furniture or other objects. All of those behaviors are the cat's way of saying, 'I'm lonely, I'm bored, I'm scared, and doing these things helps me feel better.'"[12]

It takes only one minute to keep your cat occupied and prevent loneliness –

- Purchase pet toys that hold treats so your cat – who is naturally curious – can actively hunt for hidden treats while you're away. Look for puzzle toys online and in pet stores
- Make sure he has a cat tree to play on with some height for climbing and jumping

[12]http://cattime.com/cat-facts/lifestyle/301-tips-for-leaving-cat-home-alone (Cattime.com n.d.)
[13]http://www.azquotes.com/quote/829751

- Create a nice comfy window seat so kitty can watch the natural world of outdoor activity
- Put on a DVD of animal activities or tune in to The Animal Planet network or YouTube where there are many videos intended to keep your cat entertained
- You could also adopt a second cat as a companion for kitty. See the next chapter: When kitty needs a friend. Or, in the words of Ernest Hemingway, "One cat just leads to another."[13]
- You can also install free apps on your phone that show you what kitty is doing while you are gone. There are treat-cams that dispense treats at the same time as capturing video of your cat; there are web-cams available, interactive connections for you and kitty to chat back and forth, and even cat twitter and cat book pages for the purpose of connecting while you are separated
- The last resort would be medication for your cat if he or she has longer-term issues with separation and being apart from you. A vet can help with this

Cats are creatures of habit, happier remaining in their own environment than traveling with you, unless, of course, you are *The Cat Who Went to Paris* (and rode on the shoulders of your person for decades). The 'ordinary' cat (and I realize, of course that there is no such thing as an ordinary cat), is a homebody, however, preferring his home territory to that of unknown locations, strange cats and other creatures.

One of the best things you can do is create a positive ritual for your cat to associate with your absences. When you come in the door from being away, reward him with a treat that he will remember, and come to look forward to.

The One Minute Cat Manager summary

- Buy toys that can be filled with treats and thus hold kitty's attention for longer
- Spend an entire minute connecting exclusively with him alone
- On your return, pet him, praise him, and give him a treat to let him know how very glad you are to see him again

9 When kitty needs a feline friend: adopting a second cat

If you work outside your home and have a cat, you may feel that leaving her alone for 8-10 hours a day is a cruel thing, and the thought of getting your cat a buddy may cross your mind. That happened in our household, and now there are five cats in residence. You don't have to go that crazy, but there are considerations and tips that can help with making the decision to adopt a companion for kitty.

"The domestic cat has recently passed the dog as the most popular companion animal in Europe, with many seeing a cat as an ideal pet for owners who work long hours," said Daniel Mills, Professor of Veterinary Behavioural Medicine at the University of Lincoln's School of Life Sciences.[14]

The One Minute Cat Manager has many years of experience of integrating new cats into the household. All of them have been successful and even, in one case, led to wedding bells (Percival married Harmony Rose). It takes only one minute to consider each point –

Q: How do I know if my cat wants a companion?
A: You don't. The only way to know for sure is to try it out. (Many shelters and rescue organizations will offer a 7-14 day trial period wherein you can return the cat if it doesn't work out. Some require a re-homing fee for the return. The policy varies with each country. However, as it can obviously be traumatic for any animal to be rehomed and then returned to the centre, every care should be taken to ensure the right choice is made initially.)

Q: What kind of a cat should I adopt for my cat?
A: The first consideration is health. Is the cat you are considering feral (how well might he integrate with your kitty, if so?). Spend time with the potential

[14]https://www.telegraph.co.uk/science/2016/03/14/cats-do-not-need-their-owners-scientists-conclude/

new cat to gauge how he reacts to you, and whether it appears that he might easily bond with you. Observe him interacting with other cats, and watch for aggressive behavior.

Does he have all of his shots, and are they up-to-date? (You want to protect your kitty from anything contagious. You can have your new cat vaccinated and protected, of course, but wait at least a week before taking him home to your cat, or keep him isolated until you are certain his vaccinations are in force.)

Do you know the new cat's history (did he come from a family ... how well does he play with other cats/dogs ... was he a stray)? It is important to have a sense of behavioral history, and even know the reasons why the animal is in a rescue centre.

Has he been neutered? Usually, a vet will offer discounted spay and neutering services for shelter adoptees. Plan ahead for the recovery time post-surgery as you will want your new cat to be fully healed before taking him home to meet your cat.

Is he declawed? Hopefully not, but do make sure his nails are trimmed before an introduction to your kitty.

Q: How do I integrate them?
A: There are several ways to allow the cats to meet and bond. Keep the new cat secured in a crate, and let your kitty explore and sniff the new addition. When your cat completes her explorations and demonstrates calm, open the door of the carrier and let the kitten/cat come out on his own. Show him where the litter tray is right away because he'll probably need to use it. Leave the carrier out with the door open for a day or two in case the kitten/cat feels she needs a safe space to retreat to.

You could place the new cat in a room of his own with separate litter, water and food dishes. Let the cats play footsie under the door until they have had time (a day or two) to check out each other and therefore minimise any feelings of threat or fear.

You can also put a little dab of butter or a drop or two of olive oil on the top of both cat's paws to distract and comfort them during the integration process. *The One Minute Cat Manager* loves this technique ...

Getting to know you

Q: Should I adopt a kitten that my older cat might enjoy playing with?
A: Maybe yes ... and maybe no. Yes, if your cat is under five years old. After this age, he may or may not enjoy the carryings-on of a frisky kitten. (If you've never seen a cat roll her eyes at you, this may be your opportunity.) Kittens come in all ages: perhaps a six-to-eight-month-old kitten would be just right for your cat. If your cat has lots of energy and is playful, a kitten might be the perfect companion. If your cat is more laidback than rambunctious, you might want to try an older cat with a similar personality for compatibility.

Anecdotal evidence suggests that getting a younger cat of the opposite sex tends to work out favorably.

Q: Do I need anything special for the new cat?
A: Yes. Procure another litter box and separate food and water dishes for the new kitten/cat. You'll want to make sure you feed the new arrival a few feet away from your existing resident, and by providing separate dishes and litter box you'll reduce the risk of territory issues.

You can buy special 'happy cat' pheromone plug-ins that release feel-good molecules into the air that not only soothe the cat(s) but also reduces their stress. One such brand is Feliway.

The One Minute Cat Manager suggests having three sets of toys on hand: one for old kitty; one for new kitty, plus a spare.

Tea for Two

Q: How long will it take for the cats to become friends?
A: This partly depends on the cats, and partly on your patience and understanding. Some cats may bond instantly; others may require more time for investigation and the sizing-up process. There may be some initial growling, and you'll need to watch to see if this escalates into yowling

Period of adjustment

and hissing – or flying fur. Usually, cats will have the relationship formed within a week. You have two hands: take a minute to play with each of them, one on each side of you, with different toys, simultaneously. Let both cats understand you have equal time and affection for them. You will also be creating a positive suggestion for each cat as they begin to associate fun times with being in the presence of the other cat.

Provide adequate vertical space as well. Adding a cat tree is a good idea, as well as new window seats and room for the new cat that doesn't intrude on your other cat's territory or spaces.

Be sure to make the cat introductions and allow time for adjustment before you get involved in a time-demanding assignment, leave for a vacation, or start a new project. This will set the groundwork for happy and secure feline relationships. Cats respond very well to respect and consideration of their feelings.

Two of the cats we adopted came from the same shelter, although they had not met each other. The male was a lovely, 8-month-old Lynx

Purrfect love

point Siamese mix who had a sad-sack countenance, a motley fur coat, fleas, and (as it turned out) a case of ringworm. But he possessed a personality sweeter than ambrosia. The female was a pure white, fluffy youngster barely six weeks old, feral, recently-trapped, scared and withdrawn. We were told, "Oh, you don't want that one!" so, naturally, we adopted her on the spot and took her home to be loved and re-conditioned.

We adopted both cats within a month of each other. We kept the feral separated from the rest because we wanted to show her patience and a generous amount of gentle handling – she had a tendency to bite and hiss.

Over time, the male and the female, two improbable sweethearts, fell in love. It was gradual, and the older male was respectful of the younger female's need to develop in trust and comfort. Within six months these two were inseparable: they napped together, shared a nighttime basket, wrapped their paws around each other, and were never apart for more time than it took to use the litter box. She even grew a dazzling fur coat with stripes that were very similar to his markings. It was uncanny.

He was named Percival and she Harmony Rose. They were instant friends, fell in love and got 'married.' We nicknamed them 'The Hoovers' because they both had an affinity for food and snarfelled anything they could find ... just like vacuums. Harmony died suddenly last year without warning. The vet assessed it was a heart attack, and the entire household went into shock and grieved her loss, though none as much as Percival. For weeks and months, he searched for her in each of their special places. Of course, we took much more time than a minute to comfort him, give him extra attention and space to grieve, held tightly to our hearts, but he found no consolation in anything. We began to worry about him.

Six months passed before Percival showed sign of recovering. His heart was broken. He didn't understand what had happened and we gave him the love and space he needed to mourn. He has never fully returned to his old self, but he is a sweet and loving boy who carries a vacant stare in his eye that we now believe will never go away.

Until we met and adopted this pair, we had no idea how deeply the love and soul connection can extend between felines. It is you that is caring for your felines on the physical, emotional and spiritual levels. They have needs in each area which can easily be addressed by recognizing what's going on inside the cat when he or she meets a new companion for the first time. Ask yourself, "How would I feel if I were in the same situation?" and you're on your way to the right answer for kitty.

The One Minute Cat Manager summary

- Cats have needs that are emotional, physical and spiritual
- Some cats enjoy companionship and friendship' others don't
- Prepare separate litter boxes, toys and food dishes for your new cat
- Treat them equally
- Allow time for the integration, and honor your cat's feelings about the newbie

10 Silver paws: caring for your senior cat

Inevitably, our cats will age, gradually maturing from that playful, somersaulting ball of fluff into a senior gentleman who may need assistance getting up into his favorite chair. You can help your cat enjoy a gracious silver age: growing old is not a disease, it's a privilege.

In his golden years your cat may well require some extra care and attention. Aging cats enjoy familiarity, and we can help them retain their routine and comfort as they age.

It only takes a minute to –

- Assess how your cat navigates into her favorite spots
- She may need a set of steps to climb up to her usual haunts as energy and flexibility decline with age
- Ensure your cat can also get down from higher or more awkward locations without injury
- Evaluate your cat's access to food, water and litter box. Make sure she has easy access to all three, and adapt the environment to accommodate her needs. Lower-sided litter boxes (even a baking tray), shallower food and water dishes (with fresh water daily, of course) are just a few touches that can make your cat's life easier and more comfortable
- Your kitty may urinate in the litter box and defecate outside if she has trouble with her bowels. Place newspaper around the litter box, or use training pads to absorb any potty errors
- Kidneys deteriorate in older cats, and fresh water encourages regular urination and helps to keep the kidneys working
- Observe kitty as he ascends the stairs. He may have developed arthritis in his joints, making it hard to climb, so will appreciate access to a litter box, food and water on a lower level of the house to prevent a painful

trip traveling to his necessities. You may want to make or purchase steps or ramps to help kitty access his favorite napping places. (Make sure it is carpeted to ensure secure footing and prevent him from slipping)

- Select a diet that assists your older cat with aging. Your vet will have suggestions, or you can research online for older feline diets. Stay as natural as possible. Your vet may suggest mixing together dry and wet food as chewing becomes harder with age

A leg-up might be appreciated ...

Older cats require a different balance of nutrients and calories than younger cats: be sure to choose the appropriate age-related food

In her 23rd year Lucy found it difficult to chew her food – both wet and dry. This led to experiments with the blender, some chicken stock and a shallow dish. We called these 'kitty slurpies' and made them for her twice a day. Once we started giving her the slurpies, she regained some of the weight she had lost, and had more energy

- Check your cat's teeth regularly, and ask your vet to monitor for tooth decay
- Assess your cat's energy levels. Senior cats tend to be less playful and energetic: notice whether such a change is a natural progression or if there is a health reason
- Regular checkups will provide the feedback you need
- Think warm. Your cat loves to seek out snuggly places
- Position all cat beds and nesting areas away from draughty doors and windows, but make sure the spot you choose is warm, and not hot, for senior kitty

- Aging cats may lose some, or all, of their hearing. It is kind to approach a deaf cat from the front, so that he is not startled by your presence. Be sure to keep deaf cats indoors as they cannot hear signs of danger, such as cars, or other predators
- Take an extra minute to groom and comb your senior cat. Her skin may be tender and her joints delicate so be very gentle as you comb and brush her. She may not groom herself as much as usual, so you may have to give that extra minute to help her stay well-groomed and tangle-free. Brushing helps to stimulate the skin, remove loose hair, and return luster to the coat
- Light up your cat's night. A few nightlights (not candles) placed around your home can help your kitty navigate at night especially if he is suffering from diminished vision or eyesight problems. Should your cat become

Purrfect contentment

blind, keep his environment as stationary and consistent as possible, especially food and water bowls, litter boxes and furniture. When picking up a blind cat, be sure to call his name first before approaching as this will prevent startling and alarming him

- Play with your senior cat. She may not be as energetic as she once was, but she still enjoys a slower and milder version of chase-the-treat, flying-birdie-wand, or going for a walk inside the house. Remember that she just loves being with you

- Older cats love warm laps, so be generous with yours – and your affections. Kitty loves routine and becomes set in her ways as she ages. She will also remain reassured and content if she receives tender loving care and her daily ritual stays much the same

Most of the above points take just a minute to observe and implement. The DIY project of building a set of carpeted stairs to ease kitty's climb could occupy a weekend or a trip to the local pet store. We bought ours online and they took 20 minutes to assemble.

The One Minute Cat Manager summary

- Just like us, cats slow down as they age
- Kitty may lose some or all of her hearing and sight, and experience joint stiffness and pain
- Steps, ramps, nightlights – and extra doses of TLC from you – will ease her old age

11 Feline fortitude

Lucy, the most gorgeous, coal-black cat with dazzling butter-colored eyes, was born Sept 18, 1979, inside the wall of Donna Smith's house, and rescued by an off-duty fireman.

Lucy was 22½ years old in the winter of 2002 when her persistent cough drove us to the vet's one more time. After a course of antibiotics, she was still sneezing and coughing, however, so I tucked Miss Lucy into her traveling carrier for a follow-up visit to the vet.

Lucy was very popular with her regular vet because she was his oldest, living client. He had been called away on this occasion, and a substitute was on duty. Lucy disappeared into the hands of the technicians and was taken to the x-ray room for about ten minutes, during which time I thought I heard a piercing yelp. Lucy was brought back to me and placed in her carrier by the vet. He viewed the films and a different antibiotic was prescribed. I took Lucy right home.

Lucy's usual custom arriving home was to bolt out of the carrier and head for any place that wasn't within a hundred feet of the carrier. I opened the carrier door. Nothing happened. I carefully lifted her out of the crate to discover that Lucy could neither walk nor move her hind quarters. She collapsed on the floor right in front of me. I called the vet immediately and took her straight back to the clinic.

There was a lot of tap dancing and blah blah about a 'broken femur' and 'possibly an old injury,' but I was having none of it. Lucy arrived at the practice perfectly ambulatory and left unable to walk. "She came in here for a cough and sneezing," I reiterated, "and now she can't walk." They stared at me. The vet replied, "The break showed up on the x-rays and it's probably due to cancer caused by a tumor on the bone," "Hold on!" I shrieked, "Lucy came in here walking fine and now she has a broken leg and bone cancer?" The vet nodded; I knew better – the technician had yanked on her leg and broken it.

The vet continued, "I would be happy to refer Lucy to an orthopedic surgeon who might be able to fix that leg. But," she cautioned, "Lucy's age and the probability that this is cancerous does not look good for success. I won't do the surgery here for fear she would not make it through. We just don't want to take that risk." I was in complete shock.

What began as a follow-up visit for a cough had turned into full-blown crisis. The vet was agitated and clearly wanted to be done with the entire situation to move on to her next client.

We got the referral to a clinic across town, miles away. The vet gave us a pain reliever for Lucy and bid us 'Good luck.'

Travel to the second vet took over an hour. More papers were filled out and Lucy was taken in for more x-rays.

The vet returned with Lucy and the x-rays and conveyed bad news. "We looked at these films and we believe there are four options. We are pretty sure there is a cancerous tumor at the femur joint. 1) we can amputate the leg; 2) we can perform surgery and see what bone fragments we can pin together, but it's not recommended since there's been cancer and bone loss; 3) we can just do a biopsy and wait for the results before we do anything else; 4) we could just put her down because, at her age, none of these solutions is really going to help her much. The decision is up to you." Pause. "Do you want to think about it overnight?" they asked me.

My world had been turned upside down in three brief hours. A confusing flood of emotions came over me, complex and ranging from anger and fear; to panic and sadness. Yet, under it all, was a surprisingly stable bed of conviction. It would be Lucy's decision, not mine, and I was there to support her in whatever she wanted to do. "No. I don't need any time, thank you," I said.

I looked at Lucy; she gave me one of her intense, one-minute looks, and I knew exactly what to do. There was no question in my mind. Lucy was a survivor. She had lived this long because of her strength and because of her 'Unsinkable Molly Brown' spirit. She would make the right choices for herself, and all we had to do was give her every chance to live if she wanted to.

I took the doctor aside and told her Lucy was the strongest being I knew. She was born a fighter and I could do nothing less than give her the chance to survive. I asked the doctor if she would agree to go into the OR and do her most creative work; her best job ever. "Lucy will be fine," I continued, "If you get her through the surgery, I'll take over and help her recover."

The vet didn't answer me right away; she looked up, took a deep breath and replied, "At her age surgery is a huge risk. She might not make it through. Her heart could give out. Lots of things could go wrong." I told her, "You and I know that, but Lucy doesn't. She'll be just fine if she makes up her mind to be."

The vet disappeared into the back, not entirely convinced that this was a good thing to be doing, to check on the operating schedule.

It was decided by the surgical staff that Lucy should wait until the following morning for her surgery. The doctor promised to keep her pain-free until then. I requested a speaker phone call from the operating room before they anesthetized her.

Bright and early the next morning, two surgeons, the anesthesiologist and the assisting nurse, gathered in a room by the speaker phone. I asked a few tactical questions, requested light anesthesia because of Lucy's age, and then asked the surgeons to remain positive, encouraging, and to picture Lucy's successful outcome. I asked them to encourage Lucy with healing words throughout, no matter what they found. I had a chance to tell Lucy I loved her, that everything would be fine, and I couldn't wait to take her home. I felt we were all on the same team: Team Lucy!

An hour into the surgery the phone rang. I was informed that Lucy's heart was beating irregularly, and they were worried about continuing. "Do you want to abort the surgery?" "No," I asserted, "Lucy will make the decision if she wants to go." I encouraged them to hold the faith that she would survive, and asked them to tell Lucy for me that I loved her. "Please let her know I'll make her favorite Mac & Cheese when she gets home." "Okay," said the assistant, and returned to the operating room.

The next hour seemed like the longest in history. My emotions were running the gamut and anxiety was high. The phone rang again; this time it was the surgeon telling me that Lucy had made it through surgery, but warning that they were not at all confident that the two plates and four pins they used to repair the break would hold. They were 95% sure it was bone cancer, and that it would never heal.

Lucy was to remain in intensive care for 24 hours but could go home after that. I don't think I slept a wink waiting for the next call that would release her. I knew that she had her heart set on Mac & Cheese. I bought a case of it.

I readied the car and her carrier with soft pillows and fluffy towels. The house had been prepared for her arrival with a big cushy comforter and fresh pillows in her cat bed. Arriving at the practice, I could tell Lucy was very happy to see me, even though she was groggy and in a great deal of discomfort.

Braveheart

Lucy's favorite spot to sleep was on the big purple comforter on the king-size bed, but we both slept on the floor for several nights because her leg was gingerly held in place with plates and screws, and we needed to be very cautious as it healed and mended.

Lucy couldn't eat and would only accept water through a dropper. Two days passed when the phone rang again. It was the vet calling with Lucy's biopsy report. She was dumbfounded: there was no sign of cancer – none. Lucy had beaten all the odds. Four different vets were almost convinced cancer was in her bones, but Lucy surprised us all.

The surgical team was still reeling from the fact that a cat as old as Lucy could endure the rigors of a two-hour surgery. What had been the initial prognosis of cancer and death was replaced by courage, survival and mending: Lucy was determined to live, and there was no stopping her.

Her leg healed well enough for her to walk on it, and she lived for two more glorious years. Lucy was 24 when she finally agreed to move on to her next life. She was my greatest teacher of all that is possible when you have the right mindset and a heart full of courage to back it up. In times when I have had to face surgery and healing from a critical injury, I picture Lucy's face, and summon the inner courage to go on.

Her regular vet assisted her crossing into her next life. He told me, "You squeezed every ounce of life out of her." Yep ... and Lucy squeezed right back.

The One Minute Cat Manager summary

- Never underestimate a cat's ability to heal
- Support her natural instinct to survive

"What greater gift than the love of a cat?"[15]
CHARLES DICKENS

[15]http://www.goodreads.com/quotes/tag/cats

12 Keeping kitty safe

There are many potential household hazards than can harm or injure a cat. Kittens are the most curious of all creatures, and cannot pass up a chance to jump, climb or crawl into or onto anything that piques their interest. A shiny object, a small item, a sound can catch their interest in a nanosecond. Mature cats, with environmental experience, are less inclined to act on impulse, but adult cats are not immune to danger, either.

The One Minute Cat Manager suggests you take a tour of your house, and look at it from your cat's point of view. Get down on your hands and knees, and see the world from the perspective of a kitten; remember that he is a climber and a jumper, so 'look up.' Below are listed some things to be aware of –

Potential dangers

- Open washing machines or tumble dryers. Warm clothes make a nice bed
- Ovens and hot hobs. Keep kittens off worktops for safety as well as for hygiene
- Paper shredders. Cats have curious noses. Ouch!
- Small holes that lead to the outside; chimneys, dark crawl spaces
- Ground level fridges. Kittens can be inside an open fridge door in no time flat
- Plants and cut flowers (see list of poisonous plants: http://www.petmd.com/cat/slideshows/emergency/poisonous-plants-to-cats)
- Cleaners, bleaches, disinfectants, laundry capsules and concentrated liquids. These are lethal to cats
- Automotive supplies like antifreeze, oils and cleaners

Cat cautions

- Painting supplies: thinner, stripper, paint
- Needles and thread left lying around. Very harmful if swallowed – or stepped on
- Ironing board and hot iron can be knocked over by a jumping cat, with burns and injury the result

- Wires and wiring are irresistible play toys for cats. Hide them well
- Christmas trees and decorations
- Pieces of tinsel, holly and (poisonous) mistletoe berries
- Balconies and windows in high-rise apartments are dangerous for cats (see tips below for high-rise apartment-dwellers
- Fireplaces and fire pits are dangerous for kitty. Use screens to make them inaccessible
- Open furnaces and air vents that lead to furnaces
- Open incinerators
- Broken fences with 'escape holes'

Positive steps you can take for kitty's safety

- Have your cat microchipped in case of loss
- Register your cat with national rescue organizations and pet trackers. They will help you find a lost pet
- Purchase a breakaway collar for kitty in case it catches on something and chokes your pet. Make sure there is a name tag with ID on it
- Be cautious of flea products (many are toxic), and use only what your vet recommends
- Barricade your cats from the hustle and bustle of parties, Halloween, firework displays, and other assorted rambunctious gatherings. Make sure they are safe and secure with a comfy bed and toys
- Use signage to alert guests to keep doors and windows closed so that your indoor cat does not escape
- Keep a current photo of your cat handy. If she goes missing, you will need that picture immediately. Post signs right away with your cat's photo and your phone number. An offered reward garners more attention and action
- Teach your cat to come when called. You have a greater chance of finding her if she comes when you call her name
- Shake treat jars or use familiar sounds to lure her out of hiding

High-rise and apartment window cat cautions

- Move furniture away from open windows, so that kitty isn't encouraged to climb up high
- Install a window limiter or restrictor that allows the window to only open partially, at a safe height so that kitty cannot slip through it
- Use mesh barriers or screens to cover open windows. These allow the window to fully open and remain safe for your cat
- Cat-proof your balcony if you want to let kitty play on it. There are commercially available systems to buy or make

Safeguarding against falls

The One Minute Cat Manager summary

- It takes only a few minutes of your time to assess the potential dangers that exist in your cat's environment
- A few safety measures – such as screens, gates, locks, and closed cupboards – can mean the difference between life and death for a cat
- Holidays, parties and festive occasions are perfect opportunities to secure kitty away from the crowds in a safe and familiar place
- Plants can be toxic to cats if they chew them: know which to avoid inside and out
- Research other potential hazards for additional safety

13 Famous cats; famous owners

It's no secret that cats are smart, but some of their best achievements might go unnoticed if it were not for the notoriety of their 'people.'

Ernest Hemingway

Ernest Hemingway told his wife, Mary, of his idea to build a small house for his cats close to their house, but he was worried that the cats might feel rejected and hurt if they were too far apart from the main house. He wrote to her while she was away visiting friends, and suggested that they also build a small house, closely adjacent to their house, that would be just for the cats. In the house he felt the cats could raise their 'kittneys,' as he referred to them, and have their own giant scratching post and catnip nearby.

Hemingway advised his architect that he wanted to be able to see the cats from his window whilst he was shaving. He enjoyed inviting them into the main house for meals, so there needed to be an easy access door for the cats to come inside. It is even reputed that he was drinking buddies with his cats: He drank whiskey while they drank milk.

He named his cats after famous people: Isadora Duncan, Barbara Stanwyck, Billie Holiday, Betty Grable, Joe DiMaggio, Gina Lolabrigida, Rudolph Valentino, Winston Churchill, Tennessee Williams, to name a few, as well as himself.

Legend has it that Hemingway was gifted with a six-toed (polydactyl) feline named Snow White. Snow White produced a litter of kittens carrying the polydactyl gene, and it was passed on through several generations. Since then, polydactyl kitties are often called 'Hemingway' cats.

According to *The Guinness Book of Records*, the cat with the title of most toes belongs to Jake, a ginger tabby from Canada, who boasts seven

Six-toed cat with six-toed kittens

toes on each paw, making a grand total of 28. Each toe has its own claw, pad, and bone structure.

For Ernest Hemingway, it was his cats that brought this brilliant, but often troubled, writer the greatest solace during his personal turmoil and struggles.

Abraham Lincoln

History tells us that, on the eve of the final advance of the Civil War, President Lincoln was in the telegraph hut of General Ulysses S Grant's headquarters in City Point, VA, USA.

It is recorded that President Lincoln came across three orphaned kittens. He inquired about their mother and learned from a bystander that she was deceased. Lincoln comforted the kittens, wiping tears from their eyes, and made arrangements for their care which he assigned to Colonel Bowers of Grant's staff, requesting that the motherless waifs be given plenty of milk and treated with kindness and compassion. It is reputed that Colonel Bowers promised his president that he would see to their care. Apparently, Abraham Lincoln was famous for caring for all of his citizens, great and small. Carl Sandburg's *Abraham Lincoln: The War Years, Volume IV, p146.*

Executive orders

Florence Nightingale

Florence Nightingale, The Lady of the Lamp, was the founder of modern nursing, and wrote the first textbook on the subject. She also opened the first official nurses' training program, The Nightingale School for Nurses, in 1860; she was very innovative for a woman in Victorian times.

Nurse Kitty

Florence believed that most soldiers did not die from their wounds but from contracting diseases present in hospitals during their care. One of the biggest problems with this was rats. A fellow worker gave Florence a cat, Mr Otto von Bismark, and he solved the rat problem. Soon, more cats were encouraged to spend time on the hospital grounds to dine on rats.

Nurse Nightingale maintained friendships with powerful men, but she turned down male suitors in favor of her nursing career and writing. She is reputed not to have gotten along with women, but she had a life-long love affair with her cats, and is believed to have owned more than 60 felines during her lifetime. She is quoted as saying that "Cats possess more sympathy and feeling than human beings."

After serving in the war, Florence contracted some mysterious diseases. She wrote her 'Hospital Notes,' and 'Notes on Nursing' from her convalescent room, in close company of her cats, and arranged for them to have specially-prepared meals and exacting care. Her manuscripts bear evidence of their editorial assistance: their inky pawprints.

Anne, Emily and Charlotte Bronte

Nineteenth century poets and novelists, Brontë sisters Anne, Emily and Charlotte were the daughters of a poor Irish clergyman. When their parents and older siblings died, leaving them and a brother orphaned, they were placed in cruel and abusive boarding schools, which only their vivid imaginations helped them to survive.

Later, the Brontë family fashioned themselves a reclusive life in the countryside of Victorian England where they could devote time to their writing ... and their love of cats.

Emily was painfully shy and solitary, avoiding contact with people outside of the family home, and finding solace and companionship with

Tabitha and company

her cats. It has been written that the Bronte sisters were of the belief that their cat, Tabitha, acted as a sort of spirit control or medium. Emily had trouble writing if Tabitha was staying with Charlotte, and Charlotte was not inspired to write when the cat was staying with Emily. The death of Tabitha (1843) was followed by a period of dryness, broken only when it is alleged that Anne Brontë appeared at the house with a basket full of clairvoyant penguins.

Isaac Newton

Sir Isaac Newton was one of the most brilliant scientific thinkers of all time, responsible for a variety of discoveries, including the laws of gravity and universal motion, Calculus, the anti-counterfeiting measure of putting ridges on coins, the reflecting telescope ... and the cat door.

During his studies at Cambridge University, this brilliant physicist, mathematician and scientist found that his feline companions repeatedly interrupted his studies by wanting in and out. As a solution, Newton cut two holes in the door – one for the mother cat and a smaller hole for her kittens. Apparently, it didn't occur to the genius that the kittens did not need a separate hole but could simply follow their mother in and out!

Newton never married, had few friends outside of his cats, and, in retrospect, is surmised to have had Asperger's syndrome.[16] Cats brought companionship and inspiration to the reclusive genius.

Mark Twain

Samuel Langhorne Clemens, better known by his pseudonym, Mark Twain, made famous *The Adventures of Tom Sawyer* and *The Adventures of Huckleberry Finn* (hailed as the Great American Novel).

[16]http://www.goodreads.com/quotes/tag/cats

In his time, Twain was a social commentator, an abolitionist, a supporter of women's suffrage, and a man who adored cats. He was frequently photographed with cats, and often featured them in his literature. He kept many cats at his family home in Missouri, and gave them unusual names, such as Sour Mash, Apollinaris, Zoroaster, and Blatherskite.

Twain understood the nature of cats as creatures who could never be dominated. He wrote, "When a man loves cats, I am his friend and comrade, without further introduction."[17]

Apparently, his affinity for cats began when he was a boy in Hannibal, MO, USA. It is said that, from boyhood, he was never without cats, even when he traveled. If he couldn't take them; he rented them!

Twain is said to have played an occasional game of pool with his kittens, putting kitty in a pocket and gently shooting a ball in her direction. Kitty would then bat the ball around or bat it back, depending on her mood.

Newtonian ingenuity

Even in the midst of work, Twain would stop to give deference to his kitten contingent. He always let them enter a room before him, deferring, as he claimed, to royalty. We suppose he may have even let them win at a game of billiards.

Pocket cat

[17]Ibid

Victor Hugo

Victor Hugo, infamous French writer and dramatist, whose works include *Les Miserables* and *The Hunchback of Notre Dame,* was melancholic and egotistical, but his diary was filled with references to his fondness for cats. He not only loved cats, but the first animal protection law – the Grammon Law, passed in 1850, that prohibited the public mistreatment of animals in France – became law with his support. Hugo supported the legislation and spoke out publicly against vivisection, branding it malevolent and a crime.[18]

Hugo's favorite feline was given a large red ottoman, placed in the middle of his study. Hugo remarked that "God made the cat so that man might have the pleasure of caressing the tiger."[19]

Marie Antoinette

Marie Antoinette was the wife of King Louis XVI, and Queen of France. Legend has it that whilst the people of Paris starved, she hosted lavish soirees at Versailles, and outfitted herself in expensive gowns, slippers, and diamonds. She Is remembered as having replied to a court member who told her that Parisians were starving in the streets, "let them eat cake," but this may not be the case according to other historians.

Hugo's imprint[20]

Marie Antoinette was charitable, kind and helpful; she also loved cats, and gave them free reign of the palace and gardens. Many visitors grimaced when she permitted her six white Turkish Angora cats to roam the tables, and help themselves to the delicacies thereon.

Legend has it that, during the French Revolution, the Queen made plans to flee to America, in an effort to escape the fate of her husband: public execution at the guillotine. Marie Antoinette's belongings, including her beloved cats, were loaded upon a ship, ready to whisk her to freedom. However, the royal animal lover was captured and beheaded before she could set sail.

The Queen's elegant long-haired cats departed France without her aboard Captain Samuel Clough's ship, and reached the coast of Wiscasset, Maine. After mixing and mating with the local cat population, the royal cats produced a breed that came to be called the Maine Coon, which

[18]http://www.thegreatcat.org/cats-19th-century-part-1-background/
[19]http://www.azquotes.com/quote/563570
[20]Chanoine, the cat of Victor Hugo (The British Library Board)

Let them eat tuna!

has become one of the most popular cat breeds; known for its large size, intelligence, and gentle personality. Vive la France!

14 One minute bedtime stories for cats

Most of us grew up with stories. If our parents read to us we were not only fortunate to experience treasured personal time with them, but we also had the opportunity to hear of great tales of adventure and wonder. We learned of heroes and heroines, and feats of illustrious bravery. Each story had a moral. Our imaginations soared!

Stories spark imagination, stimulate curiosity, and supply examples of how we might *be* in the world. They exist in a world of wonderment and delight.

Cats like stories, too. Many an evening I have spent reading a short story to my cats that entertains them, soothes them, and probably even mystifies them. (They do prefer a story to a song, since Mom is no professional singer.) At our house we all love kitty storytime. The following are a few examples of stories that you can read to your kitty (just a minute or two is all it takes!) that have been written by cat lovers and friends of the author: they are guaranteed to entertain you and your kitty.

Where do kittens come from?

I understand you want an answer. I can hear it in your tiny squeak of a meow, and can feel it in the way you rub furiously against the back of my legs. You want to know, once and for all, where kittens come from.

It's all relatively simple. Picture a large magic mixing bowl, much larger than that you drink out of. What ingredients should we place in the bowl? Let's start with a little drop of Energy so that kittens can run and tumble and wrestle all day long ... add a dash of Playfulness so that no one ever gets hurt. We musn't leave out Intelligence: after all, kittens are smart; certainly smarter than any mean old dog could hope to be.

Next comes a sprinkle of Independence. Remember: kittens *never* beg for food; they don't do tricks ... they certainly do not fetch. Kittens answer only to themselves. You want to sleep all day? Go for it.

Courage: that's gotta go in as all good kittens are brave; each blessed with an inner lion, roaring to bust out. We add just a slight pinch of Curiosity – not too much, because we all know what can go wrong there!

Now that we have the right ingredients – Energy, Playfulness, Intelligence, Independence, Courage, and Curiosity – we stir them round and round and round in our large magic mixing bowl – nine times we stir, to be precise, one for each lucky life – and then we add a smidge of Cuteness because everyone knows that there is nothing more adorable in the whole world than a purring kitten.

Recipe for a cat

Finally (most importantly), comes Love: oodles and oodles of unconditional Love. The ability for you to love yourself and realize what a very special animal you are. The gift for you to love and be loved in return; knowing the undeniable joy you can bring simply by curling up next to someone.

We leave the magic bowl out on the worktop overnight, and, in the morning, awaken to a curled-up bundle of sweet-smelling fur and joyfullness asleep on our bed. Is there any better way to start the day?

So now you know, my fine young furry friend, where kittens come from ...

David Congalton, author, screenwriter, and radio host in San Luis Obispo, CA, USA

(David wrote the award-winning book *Three Cats, Two Dogs: One Journey Through Multiple Pet Loss* (NewSage Press). His radio show can be heard online weekday afternoons from 3:05 to 7pm at 920kvec.com (Pacific Time)

Danu the Muse

"Your job," cooed the poet to the cat, "... is to inspire me. With your silky fine coat, sweeping tail,and fathomless, translucent jade eyes, how can you be anything but a muse? You possess all the charm, grace and elegance of anyone in the world. Having these extraordinary attributes, you *must* become my Muse."

The cat listened intently to the compliments, but yawned at the

Job description: Muse

thought of a having a career. She slipped away: work was below her. She lived on the outskirts of the human world: someplace between the galaxy and earth.

The poet had auspiciously named the cat after Danu the Celtic

Goddess of the Tuatha de Dannan – a mythical race who lived between the veils of the human and spiritual realms. Danu was renowned for her extraordinary skills in the fields of poetry, arts, science, and magic.

As the day wore on, the poet sat at his writing table, but could not give form to the feelings in his heart. Like a hairball, the words stuck in his throat and would not come out. Danu watched from a distance: finally unable to bear his frustration any longer, she moved to him in silence, brushing up against his shin gently, but with enough force to let him know she was there.

As he reached down to stroke her head, Danu imparted a spark of creativity into his hand from the sweet spot between her ears. Jumping up onto his table, she lay down in the only sunny part in prepration of sleep. Before she closed her eyes, she pushed his pen into his hand. Her work was done.

The poet wrote furiously, inspiration flowing through his veins at high speed. He wrote and wrote and wrote and wrote. When Danu awoke from her nap, she felt hunger in her tummy and meowed for her supper. The energized poet obliged her with fresh fish and broth for dinner – elegant fare that pleased the Muse. When he returned to his room, the pages of his poetry were up on their feet and dancing with each other like guests at a royal ball.

The cat was well pleased and went to sleep at the feet of the poet, who fell into his bed, exhausted. Their dreams had come true.

Kac Young
Author

When silver becomes gold

Once upon a time (which is where all good stories begin) there was a silver cat who lived in a great home, with a loving person, and a warm fireplace and plenty of good, yummy food.

This cat was smart (as all cats are). He was beautiful (of course, he was a cat, after all). He was clever (no mouse, no bird, no dog could outsmart him, not ever). He was strong (he could win the battle against the biggest adversary, he was sure). In his mind he was the most magnificent cat who had ever lived.

There were a couple of problems with this scenario, though ... He didn't exactly live in that great home (yet); in fact, the Homeless Kitten and Cat Rescue Shelter was his current abode. And he wasn't very big (quite small, actually). You could even say that he was very, very small and, and ... kind of scrawny. A lot of people did say just that, as it happened.

People would come to the Homeless Kitten and Cat Rescue Shelter

and look into the room where he was, but then shake their heads and walk away. "He's too small," they would say. "I'm looking for a BIG lapcat." Or "I need a BIG mouser for my farm, he's not much bigger than a mouse himself." "He must have been the runt of the litter, poor little thing" they would mutter and walk on by. The silver cat wasn't sure what a runt was ... but he was sure he didn't want to be one!

He tried to be big; he even practiced in front of the mirror. He would inhale deeply and puff himself up, silver fur standing on end, but the minute he exhaled his fur would lie back down and he was small again. He would stretch himself, tail to the sky, tippy-toed on all four paws, but he soon grew tired all stretched out like that, and the very second he relaxed, he was small again.

He worried and he fretted. Would no one notice how beautiful and clever and strong he was? Would no one ever see how wonderful he was? Would no one ever come and take him to his forever home? His eyes were sad and his lip seemed stuck in a perpetual pout.

Then, one day, the door to his room opened and in walked a red-haired lady. She picked up some of the other cats and held them gently for a few minutes. Then she saw the silver cat. "Oh, look at him," she said to her friend. "He is just the perfect size, and look at those sad eyes and that precious pout. If I could get a cat right now I would surely take him home. But I've just moved into the neighborhood, and what with all of the unpacking and settling-in, I couldn't give him the attention he deserves."

"No, no," thought the silver cat, "Don't go away. Don't walk out that door. don't leave me here."

But the door closed and the red-haired lady was gone.

The silver cat was crushed. His little heart felt broken he was so disappointed. He curled up in a corner and went to sleep, dreaming of his forever home and its warm fireplace.

A little while later the door opened again. The red-haired lady stepped into the room and walked right over to the silver cat and scooped him up. "I just couldn't leave you here," she purred into his ear. "I will find a way to get my unpacking done and we will settle into our forever home together."

The silver cat purred louder than he had ever purred before. He snuggled up against the red-haired lady's cheek and nuzzled and nuzzled her hair.

"We are going to be so happy together," the red-haired lady whispered. "And you are going to just love our new fireplace."

And, of course, he did!

Marlene Morris, Minister, author, founder of RelevantSpirituality.com, mother of one daughter, two sons, two adorable dogs, and five brilliant and opinionated cats

Power shift

King knew he had a good life, and took full credit for it. You see, from the time he was a small kitten he had trained his 'couple' very well. There were still a few bad habits they seemed to cling to, like upsetting King's evening nap by arriving home and immediately wanting to hold and 'love' him, and, if he wanted to be just the *teeniest* bit fussy, he would prefer that dinner arrived an hour earlier than the current schedule seemed to dictate, but, well, King recognized that these were small irritations. Meanwhile, he had every glorious day to himself after the couple left the house and disappeared, returning (only slightly late) to serve King his dinner.

Oh, how King enjoyed the days he had to himself! Mornings were spent curled up on the still-warm pillows of the couple's bed, followed by a vigorous shredding of the sofas's right arm-rest (or for variation, the curtains in the bay window), which left him the afternoon to stare moodily out of that window as animals tied to ropes, attached to what appeared to be half a couple, were paraded for King's amusement. King was not entirely sure how his couple arranged such an enjoyable show for him every day, but, because they had, King felt he had to excuse them for that strange time of the week when they stayed home for two days and became uncomfortably attentive. At least they had the good sense to provide treats as compensation!

Then the 'Very Bad Day' happened. King did not see it coming. He had assumed that the couple had disappeared for the day as they usually did, but, mid-way into King's pre-midmorning nap, the door opened, and King heard the fateful, hateful words: "King! Come see what we have for you!" Naturally, King didn't go to them. That was not his job. Hadn't he made that clear from the beginning? Before he could even finish this thought, however, the couple burst into the bedroom, and set down something on the bed in front of him. Hmmm: too big for a treat, he thought. And then it moved. And so did King. With a hiss and a flick of his tail, he dove under the bed. But matters got worse. The couple got down on the floor to try and coax King from his hiding place. "It's a kitty, King. Like you. His name is Prince. He'll keep you company when we're at work." Well! King decided, I will simply live under this bed until that *thing* is removed.

A week later the thing was still there, and thc couple were still trying to sell it to King: "Look at Prince! Isn't he cute? He's so playful!" King was not amused. And worst of all when the couple disappeared, that ... animal attempted to rough-house with King. The thing was fast, but King feigned sleep a few times and was able to bat it about the head and send it flying ... which only served to delight and encourage the interloper! King could not imagine what he had done to deserve such hell.

Begone!

Then, one day, the thing came and sat next to King on the back of the sofa. It was shaking. King was not fooled, but, as he reared back to deliver a well-placed blow, it threw up a fur-ball. With that out of the way, King tensed for the pouncing that was sure to follow, but the thing continued to shake as it stared at the fur-ball. And then, it LEANED against King. All King could think was: UGH. A FIRST TIMER. Not being heartless, King gave Prince a moment to gather himself, but the kitten simply stayed sagged against King, while maintaining eye-contact with the fur-ball. Enough is enough, King thought, and he pushed the kitten off his shoulder and made for the bedroom. The kitten slumped to the other side and, finding no support, fell over, which meant his eyes were now level with the fur-ball. He simply could not look away. The kitten felt the cushion give as King rejoined him on the sofa back. Then the kitten saw King drop a larger, dryer mass next to Prince's smaller, wetter version. The kitten looked at the larger fur-ball in wonder, then up at King. King looked back at the kitten, as if to say: "It happens to the best of us, kid."

Then, King knocked the kitten off the couch, pushed both their furballs off the back of the sofa and curled-up in Prince's warm spot. "Well," King thought, "he's lucky he has me to teach him!"

Lyla Oliver, cat-liker, supervising producer on *Madam Secretary*, lover of Martinis; @thespicybitch on Instagram

A chance encounter

Once upon a time, there was a small, gray-and-white cat named Irene. She was a very independent kitty who lived in the trees and on rooftops during the day, and, at sundown, went into the house of a certain couple she liked. She climbed up a tree, through the window, and over the top of their bed to curl up at the foot of it.

In the morning, she would allow the people to pet her a bit, but would then sail up to the window sill and fly out into the nearby tree to begin her daily neighborhood rounds.

The couple always left food and water for Irene on the floor near the foot of the bed. She liked *Meow Mix* best because it was crunchy, instead of soggy like the stuff she had been eating from the bins people kept outside their houses. She was sleek and healthy now.

Her 'people' also kept company with a big, shaggy Sheepdog named Chance, but he slept outside at night, so she had the house to herself after dark.

One night when Irene came in through the window and landed with a thump on the end of the bed, a huge, shaggy head rose up from beside the foot of the bed and said 'woof.' The couple had fallen asleep and forgotten to let the Sheepdog outside!

Irene howled. Her sleepy people bolted upright. One of them screamed, and Irene climbed the other one, using her claws to speed her access to the window. Off the sill and into the tree she soared to safety, leaving behind her one confused Sheepdog, one laughing person, and one slightly punctured and perplexed person.

Later, when she was sure the dog was outside where he was supposed to be, Irene came back through the window, took up her

Surprise visit

rightful spot, and, double-checking the foot of the bed, slept peacefully all night long.

Clearly, the moral of this story is: independent females should always know how to stay one jump ahead.

Donna Wells, proud wife of Sam and mom of Julia, and a graduate of UCLA School of Law

Donna is currently a staff attorney with the Los Angeles County Chicano Employees Assoc

The cat who turned into a bird

In the deepest and darkest part of Africa was born a brightly spotted cat, whose parents named her Swoozie because she swished and swashed when she walked.

Swoozie's parents were high-ranking members of the jungle, renowned for their hunting skills. At a very early age they taught Swoozie and her siblings how to crouch in the brambles, wait, and watch for prey. As the prey came closer and moved into range, they showed her how to leap and bound in one swooping move to pounce.

While Swoozie excelled at the graceful hunting moves, she didn't like the business of the catch, and preferred to find food in the river rather than participate in the hunt with the others. She began making excuses for her absence, or showed up late or lagged behind, and tried not to look.

One day her mother asked Swoozie if everything was alright. Big tears formed in Swoozie's eyes, and she told her mother that she didn't like to hunt the other animals and eat them for dinner: she wanted to help them, not harm them. Swoozie's mother told her, "If you want to be strong and Queen of the Jungle one day, you must be a hunter. That's what spotted cats do." But Swoozie did not like what her mother said, and quietly made her way back to the river.

At the river Swoozie had friends. She liked to play with the jungle squirrels; the monkeys always made her laugh, and the happy songs of the tree birds brightened her day. Swoozie was happiest at the river. "Leave the hunting and grabbing to the rest of them," she thought to herself, "I'll stay right here."

One day a big storm blew up. The jungle was shaken by winds; rain came down like bullets, and the river swelled, carrying everyone in it far downstream. All of the animals were hiding under whatever cover they could find, and, from her branch-covered hideout, Swoozie heard the frantic peeping of baby birds. The sounds grew in intensity and volume until she had to cover her ears with her paws.

The stiff winds had knocked all of the nests to the ground and the babies were scared and calling out for their mothers. Swoozie knew that the mother and father birds had been blown out of the trees, too. She did not know if they were ever coming back.

One by one, Swoozie gathered up the nests and baby birds and took them into a nearby cave. All of them were hungry and thirsty, so Swoozie scratched the bark of the trees with her long, sharp nails and found some bugs to give to the babies. Then she found pointy leaves, filled them with water, and angled them into the mouths of the babies. After hours in the kitchen, Swoozie fell asleep surrounded by the babies.

One of Us

At dawn, the babies woke her, and the feeding routine began again. Swoozie didn't mind: she knew just what to do and fed the babies another delicious meal washed down with fresh river water.

One by one the frightened parents began to fly back to the trees. Most of them were afraid their babies had died in the storm, and they were astonished to see Swoozie the Jungle Cat feeding their children ... they would have expected her to eat them!

The parent birds were delighted. They called a brief meeting among themselves, and unanimously decided to camp in the cave with Swoozie and the babies: there was no sense building new nests again this late in the season. The father birds stood guard during the day, while the mother birds hunted for food, then they stood guard during the night while the fathers kept the babies warm. Swoozie continued as water-supplier by leaf ... and that's how she became known as Swoozie Nightingale, and initiated as an honorary bird by the entire flock. They gave her a medal that read, 'One of Us.'

To this day there is a monument at the entrance to the jungle dedicated to Swoozie Nightingale, the cat who became a bird.

Kac Young
Author

Cats in London

It was a dark and stormy night – much like this one – when Betty and Carmello snugged down into bed. Just then, the tiny attic apartment was stabbed by light. A heartbeat later, thunder rumbled.

"What does this remind you of, Carmello?"

A bump in the covers moved and a chocolate-colored Burmese poked his head out of the sheets. Through long, Wild West sheriff-looking whiskers he said, "it reminds me of our greatest case ..."

Tiny Socks, the new kitten, was missing. She had arrived three weeks ago, and spent her time sniffing about. Her children-people, ecstatic to bring her home, were now crying and refusing to eat or sleep. Foul play was suspected.

Search parties were formed, and Betty led the humans, floor-by-floor, while Carmello called up the others – Raffi, Cezanne, Toots and Frank – to search cat places. Rizzo, Rinky and Rapunzel searched the rat spots, and even old Bubby got involved ... that dog hadn't moved from the spot by the door in a decade! Bubby had bad breath and broke wind often, but his tail-thump greeting made him beloved.

For 24 hours they raced and wiggled through walls and cupboards; pipes and drawers. Downstairs, Bubby watched all the activity and let out elderly barks. Rinky found Mrs Oliver's blush underthings: she hadn't worn them since the honeymoon! She sat down staring at the cloth, eyes wet and smiling. Cezanne was calling "Socks! Socks!" under Mr Windleston's sofa, and saw a light winking in the dark. It was Mrs Windleston's ring, lost for a generation. Rinky swore he saw Mrs Windleston put it on, grab her husband and dance. No one had ever seen her without a shopping bag or a frown.

After a long day of investigating human places, becoming small and flat to slide into or through wherever Socks might hide, the searchers ended up in the front hall. As if hearing a call inside their heads, first Toots and Frank showed up, then Raffi and Rinky, and finally Cezanne and Rapunzel, fur matted, ears back, panting and in need of a break. Betty arrived, completely frustrated. She sat on the stair; Carmello by her side.

"We don't understand the mind of this cat," Betty said. "What do we know?"

"Precious little," sneezed Carmello (all over his mistress). "This little cat has disappeared without a trace."

"I remember the times I saw her – I had to run away quickly ... mind you, I'm a rat – she always had her nose in the air, sniffing," Rapunzel said.

"She was homesick," Frank said. "She spoke of it often."

Betty leaned forward, furrowed her brow, and said, "So we do know a bit. She's searching for the smell of home."

The foyer fell quiet, remembering those feelings of first arrival in London; the excitement, loneliness, and fear.

"If you're homesick, where do you go?" Betsy asked.

"I sleep in the sun!" shouted Cezanne.

"I nestle down in the covers by my Mistress' side," murmured Carmello.

Rizzo rocked forward on the balls of feet and said, "Rolling into a ball in a drawer is best."

Raffi burrowed into the sitting room pillows, Toots preferred a shelf above the heater; Frank was a cuddler, surfing human heat, and Rinky and Rupunzel agreed that a dark passageway and a wad of hair made them feel happy and safe.

Had someone taken Socks? Had she run away? Each of them had a first place, a home with real parents who loved them well. Through forces out of their control, they were in a city with strangers. Dark images of Jack the Whisker Nipper flitted through their heads. How many times had they just gotten across the street when the lorry rocked past? Why were humans always rushing? Why did they scream when they saw Rinky, Rapunzel or Rizzo, or kick at Toots and Frank? This world was confusing.

Brains buzzing with questions, it took a moment longer than usual for their hearing to pick up a high-pitched sound bouncing around the room. It came in short insistent bursts, in upper register, a question in need of answering.

"What's that?" Betty felt a jolt of hope.

"We hear it too!" rasped Rinky, Rizzo and Rapunzel as Toots, Frank, Raffi and Cezanne fell into a scrum of face-rubbing and tail-twitching. "We think it's coming from there!" Three boney little rat fingers shot up and pointed at the pile of fur, just inside the door, rising and falling as Bubby snoozed through the cries of his excited housemates.

Carmello leapt from his spot by Betty's feet and shot to the large dog, lying half on the door jam, half into the garden. He sniffed the beast just as Bubby's head came up, eyes opening. His tail thumped. A breeze kicked up, grabbed Bubby's escaping aroma, and distributed molecules over the searchers.

"Pff!... awk!....." filled the foyer as everyone shifted to escape Bubby's unique perfume. Rinky heard something more; a tiny 'mew' punctuating the 'yucks' and 'pee-yews.'

"She's there!" shouted Rinky, as Carmello searched Bubby's fur, finally locating a tiny black-and-white nose, popping out of dog belly.

"Bubby," Carmello said softly.

The tiny ball slid to the floor, unrolled, and began to clean herself.

Bubby Love

"Socks!" They shouted at once. She looked up, shocked.

"Wha ...?"

"We thought you'd been taken!"

Bubby's tail beat the floor faster and Socks smiled. "He smelt so good – just like the barn that was my first home – he smelt good and I just wanted to snuggle." Bubby's tail thumped. He lowered his large head and licked her as the children-people tumbled through the door, school over.

"I remember the joy in the house that day," Betty mused as Carmello rearranged himself behind her knee. "I expected to find an evil animal catcher or careless owner at the center of this mystery."

"Bubby saved the day," yawned Carmello.

"Yes, it was our greatest case. The dog saved a cat. Who would suspect?" They snuggled deeper, warm with the knowledge that, in a loving home, anything was possible.

Beth Wareham writer, publisher and mistress of LaLa, a little Bombay who thinks she's a female rapper, and Carmello, a dreamy gentleman of a Burmese, who loves a patch of sun

Mistaken identity

Barney was an ordinary cat (if there is such a thing) who lived in a house with four people – Dad and Mom and Sissy and Doug – and two dogs.

The dogs were very different from Barney. Barney was young and small and gray. Bianca, the oldest dog, was tall and red and very long-legged, which allowed her to run very fast and jump over just about everything. Snickers was short and round and fluffy and white (when he wasn't dirty – which he often was). He had a funny side-to-side, Charlie Chaplin kind of walk, and a tail that was always wagging, especially when he was being fed, or petted (which was most of the time).

The dogs, Barney decided, got more attention and had a lot more fun. They got to go outside and play. They jumped up and caught things. They went for walks with Sissy and Doug. They slept next to Mom by the warm fireplace and got petted. They got to go to 'the groomer' (whatever that was). And their dinner dishes were a whole lot bigger than Barney's.

Barney longed to be a dog, and so the tiny cat formed a big plan in his furry little head. He would watch the dogs very carefully, and then he would mimic everything they did. If he could act like a dog, he reasoned, his people would soon forget that he was just a cat, and they would take him outside to play and let him sleep next to them by the fireplace, and pet him, and he would get to go to 'the groomer.'

And, best of all, he would get a bigger dish for his food.

Barney watched Snickers and Bianca very carefully for the next week and tried to learn. The dogs both ran to the door and jumped up and down when Dad came home, and Dad smiled and petted them and said, "Good dogs!" Bianca was outside playing with Sissy and Doug. The children would throw a dish-shaped thing and Bianca would jump up and catch it and bring it back to them. The dog was having lots of fun, and Sissy and Doug laughed very hard.

Snickers went outside and rolled in the grass after the rain. He came back into the house muddy and happy and wagged his tail. Mom said, "You'll have to go to the groomer now," and put Snickers in the car and drove away. When they returned, Snickers had a blue bandana around his neck and smelled really good. Mom gave him a treat and a kiss.

"I can do it," Barney thought, "I can be like a dog," and he set about putting his plan into action.

The next day when he heard Dad's car in the driveway, Barney raced to the door and, taking a gigantic leap, jumped up to greet him. But he landed on Dad's chest and his little claws got caught in Dad's sweater and pulled out several threads. Dad scolded, him: "No, Barney, get down!" and, plucking the cat's claws from his favorite golf sweater, unceremoniously put him back down on the ground.

"Well," thought Barney, "that didn't go so well. But I'll do better next time."

Later that day, Barney thought he saw an opportunity. The children had left the dish-shaped thing on the kitchen counter and, flinging caution to the wind, the cat jumped all the way up to the counter, but he overshot his mark and landed in a bowl of bread dough. Hearing the noise, Mom came into the kitchen and saw him, covered with dough and drying quickly. "I can't get all this sticky dough off," she said, "You'll have to go to the groomer," and she scooped up Barney and put him in the car.

"Oh boy" thought Barney "I'm going to the groomer, just like the doggies. This is going to be great!"

What happened after that we can only imagine ... but we do know that it involved Barney getting very wet, and very soapy, and then very wet again, and then having a lot of air blown in his face. When they were done they tied a pink ribbon on his head, and Barney didn't like any of it; not at all.

That night Barney sulked in a corner of the living room. He had really failed at being a dog. He had ruined Dad's sweater, he had spilled the dough, and he hated going to the groomer. His dream was dashed, his pride was hurt, and his hair was disheveled from unsuccessfully trying to remove the pink bow.

"Poor Barney," Mom said as she scooped him up."you've had a bad day. Come and sit with me by the fire."

Mom placed Barney on her lap and petted his little gray head."You are a great cat," she cooed as she removed the offending pink bow,"and we love you.You've been acting a little strangely today, but I guess that's because you are a cat and cats like to get into things. But when you are done getting into things you can always come here and sit on my lap and we'll be warm and snuggly together." She kissed him and then she said,"Dad, Barney is really growing. Don't you think we should get him a bigger food dish?"

Barney meowed really loudly, and it made everyone laugh. He looked over the edge of the big chair and saw the dogs laying on the floor below him, and thought,"Maybe it's good to be a cat after all.And maybe, just maybe, its best to let others be who they are and just be the best you that you can be."

Marlene Morris, Minister, author, founder of RelevantSpirituality.com, mother of one daughter, two sons, two adorable dogs, and five brilliant and opinionated cats

A squirrel named Merle

Once upon a time there was a squirrel named Merle, and he had a lovely girlfriend by the name of Pearl.They lived together in a hollow tree, and were surrounded by peace and tranquility.

One day out walking the pair overheard someone talking about a dance, being held – by chance – right around the corner. Pearl said to Merle, "Let's give it a whirl!" so off they went in search of the event.

It just so happened that little Miss Pearl had a magnificent tail that she loved to twirl, and, as they stepped out on the floor and started to swirl, the music began to soar, and they gave the other dancers a real what-for.

Though most of the others greeted them with a smile, some of the dancers didn't like their style.They made fun of the couple and made them cry: Merle and Pearl didn't understand why.

Then the people turned mean and created a scene, making fun of Pearl's looks."You don't belong here!" they said with a sneer and Merle stepped up to protect his Queen.

"What's wrong with my Pearl?" said the brave little squirrel."She's a wonderful girl!" But the crowd started to laugh and then threw some rocks and said "She's not a squirrel, you dummy – she's a fox! Can't you see that you two are not alike? Why don't you both just take a hike?"

Pearl was sad and Merle got mad, and they walked on back toward

Opposites attract

the home they had. Pretty soon they crossed paths with a cat who asked them why they seemed sad like that. They told the story of the unhappy dance, and the things the mean squirrels had said by chance.

"That's just silly!" laughed the big orange cat. "It doesn't matter if Pearl is a squirrel or Merle is a bird, what matters is you're in the same herd. Life puts us together with all kinds of others, and we need to get along like sisters and brothers! You picked each other to go through life, so hopefully you'll soon have a fox for a wife."

Merle and Pearl thanked the very smart cat and wended their way back to their little flat. They learned a great lesson at the dance that day: don't let anyone tell you how to play, or what to do, or who to woo – just do things your way!

Pamela Ventura, Pilates teacher, writer, trapeze aficionado, delighted wife of Ralph and abject slave of Abner (a ginger male), Little Mama and Morticia (mother and daughter Torties)

Captain Catnip vs the Aliens

Captain Catnip wasn't just your typical everyday cat. He was a *space* cat.

Captain Catnip also had a boy of his own. The boy was a small-sized human, and he built their spaceship, which was called the USS Adventure. It was made from cardboard boxes (as any cat can tell you, few things in the universe are as much fun as cardboard box), and they also make really good spaceships when you need one.

Sometimes, Captain Catnip and his boy would take their spaceship for adventures across the Milky Way. They never really went anywhere, of course, but Captain Catnip didn't mind. The boy would make funny noises like 'Vroom!' and 'Pew! Pew pew!' and sometimes Captain Catnip would make noises with him. The most fun thing about exploring the Milky Way, though, was that when they got there, the boy would say he was going to explore the planet and walk to the kitchen. When he came back, he always had a bowl of milk for Captain Catnip.

Ready for take-off!

Today, Captain Catnip and his boy were travelling to a planet they'd never been to before. The boy said they needed to be careful, because aliens lived there, and they might attack. When they landed, the boy told Captain Catnip to stay with the ship while he checked things out. Then he walked over to the box where he kept Captain Catnip's toys and grabbed a bunch of them.

The boy yelled "alien attack!" and threw the toys all over the spaceship. There were little furry toy aliens, and toy ball aliens with bells in them, and even one alien that smelled like catnip. Captain Catnip leapt and swatted away one of the alien balls, and the bell jingled as it rolled across the floor. Then he turned and jumped back and batted across the room an alien that looked like a mouse. Then he pounced on one that looked like a little ball of fur. The boy laughed, and Captain Catnip had to admit that this was the most fun alien attack he'd ever seen.

After a while, it was time to return to Earth. The boy cleaned up the alien toys and put them back in the box, and gave Captain Catnip a scratch behind his ears.

"You're the best co-pilot ever," the boy said. "I can't wait to see where we go next."

Patrick Morris, husband of RaeAnn, computer geek, and part-time amateur zookeeper

Jebby goes to the beach

"Where are you going with my water dish?" Jebby purred, rubbing his somewhat shaggy gray body against his Mommy's robe. "Why are you putting it in that big blue thing?"

Jebby was a lovely cat who had a tendency to look like a bad fur coat when he stayed up too late; which he had done the night before. Mommy had company and they stayed late, and Jebby didn't want to miss a word of the conversation.

"This is a cooler" Mom said, as if replying to his question. "We're going to go to the beach today. Look, Jebby, I'm even putting some of your treats in a bag to take with us to the beach."

Jebby jumped up on the counter (which he knew he wasn't supposed to do) to get a better look inside the 'cooler.'

"Jabez, get down!" Mommy yelled. (Jabez was Jebby's real name but Mommy only used it on special occasions.) Just before he leaped he thought he saw a round of cheese, a bottle of red stuff, two glasses, and his water dish.

"I like cheese; I love to go out with Mommy and I love my treats" he thought, as he landed somewhat unceremoniously on the floor, "Whatever a beach is, it can't be too bad."

As if in answer to his thoughts, Mommy said, "A beach is like a great big giant litter box, Jebby. You will love it."

As Mommy left the room to 'get ready,' Jebby curled up in his bed and pictured his litter box filled to the top, and then tried to imagine it bigger and bigger. He smiled his kitty smile and purred very loudly.

In a little while Mommy was back. She was wearing a sundress, flip-flops, and a big hat. She picked up Jebby and put on his harness, then plopped him into a large bag, put it over her shoulder, and took him to the car. Jebby saw his purple stroller in the back seat next to the blue 'cooler,' and got very excited.

Jebby loved to ride in his purple stroller. He had such a good time when Mommy took him to the airport in it. Everyone gathered around and said, "Is that a cat in your stroller?" And Mommy said "Yes, that's Jabez. He loves to travel." And everyone ooohed and aahhed that a cat would actually sit still and travel in a purple stroller. And Jebby smiled his kitty smile and purred very loudly.

They drove for quite a while and Jebby began to get nervous. "Are we there yet?" he meowed. "How much farther is this beach?"

"Just a few more minutes." Mommy said. "We're almost there."

Mommy was right. In just a few minutes she pulled the car into a parking place and turned off the engine. Her friend jumped out of the car next to her, gave her a hug and said "I'm so glad you both are here. Can I help with anything?" While they were getting things out of the car, Jebby put his front paws on the dashboard and stretched so that he could see.

He almost fell off the seat. There, in front of him, was the biggest litter box he had ever seen. He looked to the left and litter was all there was. He looked to the right and it was lovely golden litter as far as he could see. And straight ahead there was litter all the way to the water ... which was more water than he had ever seen in his whole kitty life.

Mommy reached in and lifted him and put him in his stroller; Jebby rode in his purple stroller out onto the big litter box.

Mommy's friend put down a blanket and Mommy put the cooler on it and they sat down. She took Jebby out of the stroller and put him on the blanket; then she opened the cooler and brought out the bottle of red stuff and poured a little into each glass. Jebby sniffed the glass but he didn't much like the smell of the red stuff. His nose curled and he meowed his displeasure. Mommy poured some water in his water dish and said "Cheers, Jebby, welcome to the beach."

Jebby was getting hot and the water was welcome. In a few minutes Mommy said, "Let's all go for a walk."

That's what he had been waiting for. "Oh boy," Jebby thought, "finally I get to go into the big litter box."

Mommy checked his harness and put him into her bag. Her friend opened up a brightly-colored thing she called a parasol, and held it over the gray cat because it was "getting pretty hot."

"No kidding" Jebby meowed. It was hot, and he wanted out of Mommy's bag.

Then it happened: Mommy reached into the bag, picked out Jebby, double-checked his harness, and placed him on the beach.

Jebby was too shocked to meow, but Mommy saw the look on his face.

"It's sand, Jebby. The beach is made of sand. Come, let's walk a little."

They went a few feet, Jebby stepping high on his front legs and then shaking his paws feverishly, trying to get the 'sand' from between his toes.

"I don't think he's very happy" Mommy's friend said. "Neither do I" said Mommy. "Maybe this wasn't the best idea."

"This stuff is hot and gritty, and I'll never get it off me" Jebby thought. "Yuck!"

Jebby dug in his heels, sat down in the offending sand, and wouldn't go one foot further.

Mommy picked him up and

Sand is a place for a clam, not a cat

tried to soothe his angst. In a few minutes he was back in the car and on his way home. Jebby watched out the back window until he was sure that the 'beach' was way out of sight.

Mommy had placed a couple of treats in Jebby's bed as a peace offering, and he munched on these as he began to fall asleep. "What a terrible experience that was," he thought as he shook grains of sand from his left paw. "But there's one thing I've learned: the beach is no place for a cat, and I'll never complain about the size of my litter box ever again!"

Marlene Morris, Minister, author, founder of RelevantSpirituality.com, mother of one daughter, two sons, two adorable dogs, and five brilliant and opinionated cats

How Bartholomeow got his name

Once upon a time there lived a kitty named Bartholomeow, or Bart, for short. One stormy night, cold, wet and shivering, Bart arrived outside a castle. The thunder and lightning scared him, so Bart began to meow, softly at first, then very loudly, hoping one of the guards would take pity on him and let him in.

Soon, a knight passed by on a magnificent horse and, with his long, strong arms, scooped up Bartholomeow, wrapped him in his cloak, and off they rode across the moat and through the castle gate. The knight knocked thrice on the kitchen door, paid the scullery maid sixpence, and bid her take in Bart, and give him something to eat and a bed for the night. Bartholomeow happily ate some fish and lapped the milk the maid poured into a saucer for him. As he curled up on a rug near the hearth, the maid told him, "Tis only for the night, kitty. In the morning thou must leave, for the cook doesn't like cats." "Doesn't like cats?" Bart said to himself. "Well, she's never met the likes of me."

Before dawn Bart was up and prowling about the kitchen. He caught five mice, which he laid out in a row, then groomed himself and proudly stood by his catch. When the cook arrived and saw the mice Bart had caught, she clapped her hands together and said, "My, my, you've earned your keep, kitty. I dub thee 'Chief Mouser of the Castle.'" The servants all laughed, but Bart wore his new title with distinction.

As Bartholomew patrolled the castle he saw his shadow and ran down the stairs. When he went by the great hall and saw his reflection in the mirror he jumped straight up in the air. When a servant dropped the silver and it fell with a clash he crept behind a curtain. Oh, if only he wasn't such a scaredy cat! He wanted to be brave like the knights he saw riding by on their horses, and strong like the knight who saved him.

One day as Bart was roaming the castle in search of rodents, he overheard a Lord plotting to overthrow the King! Bart had to warn the King ... but how?

That evening, when the King and all his court were in the Great Hall, Bart quietly dragged the biggest rat he could find up the stairs. He crept past the mirror, keeping low so that he wouldn't see his reflection, and laid the rat at the feet of the Lord who was plotting against the King. Deep in the crowd someone called out "I smell a rat!" Just then, the Lord who was plotting against the King bolted from the room, and the guards quickly

The grateful King walked over to Bartholomeow and drew his sword: Bartholomeow quivered at the sight of it, but then, gently, the King touched Bart's shoulders with the sword, and said, "You've saved the kingdom. For your courage, kitty, I dub thee Bartholomeow the Brave."

Bartholomeow the Brave

And, Bartholomeow, the King, and his Queen lived happily ever after.

Colleen S Kennedy, PhD, Ziggy's mom and Emeritus Dean & Professor, College of Education, University of South Florida, USA

Peevorotti sings

Once upon a kitty time, not so long ago, there was a very big, very beautiful, brown-and-beige-and-white cat whose name was Peevarotti.

Peevarotti was a fabulous feline, as all cats are. He loved to cuddle; he loved to play; he loved to chase moving lights and catch toy mousies, and he really loved to eat (which is probably why he was a very big cat). But, more than anything, Peevarotti loved to sing.

Mostly, Peevarotti sang opera. He could sing in a very low voice (me-he-he-he-yow), and he could sing in a very high voice (meeeeeeee-yow). He had an amazing vocal range. Most nights (Peevarotti tended to sing mostly late at night when there wasn't so much daytime noise to distract him) he would go around the house singing his songs, going from high notes to low notes and back again, for the entertainment of his entire family.

There may have been some, however, who did not appreciate his warblings, and referred to them as 'caterwauling,' and maybe others who

O solo meow!

didn't know opera well enough to recognize the arias, the very long songs, he was singing, but Peevarotti didn't care. He sang because he loved to sing, and it made him happy.

In spite of the other cats' grumbling, some less-than-flattering reviews, and occasional complaints from his people, Peevarotti sings every night. He strolls around the house, up and down the stairs, on and off the beds, round and round the furniture. He sings very high notes and he sings very low notes, and everything in-between.

Every night, Peevarotti sings very loud and very long – and it makes him very, very happy.

My dear kitty, learn from Peevarotti. Do what your heart desires. If you want to dance, dance. If you want to scratch, scratch (but be sure and use a scratching post and not your people's furniture; that tends to make them cranky). And if, at any time, you want to sing, sing loud and long. It will make you very, very happy.

Marlene Morris, Minister, author, founder of RelevantSpirituality.com, mother of one daughter, two sons, two adorable dogs, and five brilliant and opinionated cats

Sometimes, a cat needs a longer story. It could be on a road trip, after a stay at the vet's, or when you've been away a very long time. This next tale will give you a lovely fifteen-to-twenty minutes of story time with your loving cat.

Smoky

Smoky walked down the sidewalk, upright. with a regal stride, his grey-and-black tail switching back and forth like a town sheriff holding a walking stick while on patrol. He had been strolling this block and the one next to it for so many years that Smoky couldn't recall ever being anywhere else. A warm wind blew on his face making his whiskers twitch. Something off to the side sent Smoky's sharp eyes searching. ... they came to rest on a paper bag blowing over the grass near the curb.

Smoky didn't like paper bags. He never admitted to himself that they scared him, but whenever he saw one he always considered whether or not it was a threat. This bag (which once had a pack of cigarettes in it), Smoky assessed, was too small to cause anyone trouble. Dismissing the bag he saw a bigger problem heading towards him. A little boy, about two- or three-years-old, was running in Smoky's direction little legs and feet moving as fast as they could. Just before he reached him, however, Smoky jumped up onto the retaining wall next to the vine-covered old house. Already safe from little hands, Smoky took no chances and plunged into the purple branches of the nearby bougainvillea, from which vantage point Smoky noticed a woman with a stroller reading a sign on a pole (people were always reading posts on that pole). Smoky noted that he had seen her before: she was the mother of the little boy. The mom looked up and called out, "Kyle, wait for Mommy!"

Smoky thought the woman looked scared. He had seen that look before on the faces of parents afraid their kids would run into the street. She raced and caught up with the boy and put him in the stroller. The boy turned around and pointed to the bushes and said something to the mom. She turned and looked, and Smoky thought she could see him. But then she turned around and the two of them went off down the block and into the distance. The bougainvillea was warm from the sun, and safe from the curious hands and quick little feet of small children: Smoky settled into a nap.

That evening, after Smoky had scavenged chicken bones out of a trashcan, he was cleaning his face when he thought he heard a dog. He had been around dogs plenty of times – even ones that sounded strange, like this one. Smoky moved into defense mode by jumping onto the roof of the nearest car. On top he crouched low, knowing that he blended into the night sky. Hidden from canine eyes, he thought he heard snarling across the street where Juniper, a big old Maine Coon, lived. Some kind of tussle was going on, complete with loud screeches, but Smoky had seen this before and Juniper always knew how to take care of himself. Smoky settled back into cleaning his face, using one paw at a time.

The next morning Smoky woke to hear a lady crying: it was the woman who lived in the house with Juniper. She had seen something that had upset her, but Smoky was too far away to see what it was; he thought maybe someone had broken into her car. That happened *a lot* in this neighborhood!

A few days later Smoky slipped under a car, avoiding the pink puddle where the radiator dripped. He had seen plenty of younger and more naïve cats lick their fur after being dripped upon by this stuff; they later crawled away to some place and died. Smoky was too sharp to suffer death by radiator fluid. As a keen observer of neighborhood comings-and-goings, Smoky knew that this car was there for the day; safe for a little midday nap. He curled up into a tight ball next to the right back tire and closed his eyes.

It was the pungent smell of tuna that woke Smoky, and he could tell by his shadow that he must have been asleep for a while (he noticed he was sleeping more, which came with older age). Smoky stretched his tired bones and peeked out from under the car. Sure enough, there, in the shade of a tree, was a can of tuna.

Smoky approached the can. Inviting as the smell was, he was concerned, however, because the tuna appeared to be surrounded by a box. He had seen trash – some of it very tasty – in a box before, but this one looked strange since the box was on its side. He proceeded with great care, and at each step he assessed if the smell was worth the concern of the box. The closer he got, the more his stomach told him to go for it.

He placed a paw inside the box, then another, then another, and another, and – as he took his first bite of the delicious tuna – he heard a slam ... which was when Smoky realized that this was not a box, after all, but a trap!

Smoky screeched horribly, and the next thing he knew the box was in the air and he was being carried, though he could not see who was holding the box. The tuna sloshed out of the can and onto his coat. Smoky could not believe what was happening to him – a messy coat and he was trapped! He was carried across the street and down the block past Juniper's house, and then across a street and another ... Smoky thought he had never before been this far from home.

Finally, after turning down a driveway, Smoky was carried up a couple stairs and into a porch, and place inside on a carpet. Smoky didn't like this at all. He was an outside cat, after all, and didn't want to be inside someone's porch. The box was opened and Smoky stepped out to see the woman from a few days earlier. The little boy was not there, but she was talking in a gentle voice.

"Let's make sure you are safe, Smoky."

He was not surprised she knew his name: everyone did in the neighborhood.

"Here's a water bowl and here's a food bowl." She pointed to a corner in the porch, and then added, "There's a box with sand in it; you'll figure out what to do."

The lady left the porch.

Smoky hid under the couch all day, too afraid to venture out, and the woman checked on him. Then a man, who carried the little boy, checked on him. The little boy said, "Smoky" but the dad would not let him get down. Smoky watched them all but would not move from under the furniture. Later that night, when everyone was asleep and the house and backyard were quiet and dark, Smoky walked over to the water and took a sip, then ate a bite of food. It was good, and once he felt a little return of energy from the dinner, he climbed the screen door and pulled on it, hoping to open it, but it would not budge. He tried to push open the window but this, too, would not move. Resigned to the thnought that an exit plan would have to wait, Smoky went back to the food and ate some more, then jumped up onto the couch. It was really soft, so he curled up and fell fast asleep.

The next morning the boy banged on the glass door between the porch and the main house. Smoky stared at him from a table behind a chair (he had moved from the couch to the table early in the morning when the sun peeked above the trees).

The woman pulled the boy away from the door, "Kyle, stop banging on the door; you will bother Smoky. We need to be gentle." She came into the room and put down a new dish of food and fresh water. Although Smoky didn't like the porch, and didn't like feeling trapped, there was something about the lady that didn't scare him.

Later in the day, Smoky realized he had not gone to the bathroom in what seemed like an eternity, so found a spot in the corner of the porch near a bookcase to pee. The lady didn't notice this until the next day, when she looked at the box with sand in it.

"Hmmm, Smoky, you have not used the box, I wonder where you have done your business?" Smoky watched her searching behind the furniture, sniffing quite loudly. Reaching the bookcase she said, "Got it." Leaving the porch, she came back with a mop and cleaned the spot.

For the next week the lady brought fresh food twice a day, and filled the water bowl and cleaned the areas where Smoky peed and pooped. One day, Smoky became curious about the box with sand in it, and placed first one paw and then another paw inside. The grains felt good on his paw pads. He didn't know why but he felt compelled to dig in the sand with his paws, and scratch, scratch, scratch ... and then he peed. When the lady saw what he had done she said, "Oh, great news, Smoky, what a good boy you are." Smoky didn't know exactly what it was she said to him, but there was something about the way she said 'good boy' that made him want to purr.

The lady made daily trips to the porch for a month, and the little boy sometimes came with her, but she made sure he left Smoky alone. "Let him be, Kyle, we want him to feel safe with us." And so this routine continued: the lady came in and the food, water and box litter were renewed, she said a few words to Smoky, and then she left. This pattern went on all summer until one day in fall, when the air was colder and Smoky was extra-hungry. This time, when the lady entered the room, Smoky looked up at her and meowed. "Smoky, how nice of you to greet me" she exclaimed, and, with that, scratched Smoky behind the ears.

Smoky liked the feeling of being scratched: it reminded him of something, but he couldn't remember what. The season continued, and every couple of times that the lady brought his food, Smoky would meow and run to her, and she would give him a good scratch. It got to a point where Smoky thought he liked the scratching more than the food.

One very cold the lady came in and scratched Smoky behind the ears, but did not offer food. Instead, she opened the door between the house and the porch, and placed a bowl of food on the floor of the house. At first, Smoky was not sure about the food but the lady was so nice, "Smoky, can't you trust me now?" she asked, propping open the door and walking away.

Smoky looked around: the boy was nowhere to be seen. Smoky crouched low to the ground, something he had not done in months, and darted through the door, aiming for a spot below a chair. The house was warmer than the porch; it felt good to his old bones. Smoky waited there a long time, and when the lady was busy in kitchen, raced for the food, wolfing it down, then returning to his spot under the sofa to snooze.

A few hours later the boy came running into the house. He called,

"Smoky, where are you?" Smoky had moved from the den to under the bed in the bedroom. When the boy found him, Smoky hissed. "Oh, Mommy, Smoky hissed at me just like you said," he told his Mommy.

The lady placed a soft cat bed in the corner of the bedroom. "You are going to have to give him time," she told the child.

Later that night when everyone was asleep, Smoky toured the house. He found that his food had been moved to the kitchen, and his litter box was now in a small room where the mops lived. Returning to the bedroom, Smoky decided to get into the cat bed on the floor. It was soft and cozy, and felt good against his skin. In the middle of the night Smoky woke and heard rain pattering against the window: it felt good not being outside getting wet. He went back to sleep.

For the next few weeks Smoky hid during the day and came out at night. A pattern had become established: the lady would take the boy out of the house in the morning, then come back and do things in the kitchen and clean the house. She sometimes went into the room where the box was located, where there were two machines that made a lot of noise. One day Smoky jumped on one of the machines and it was very warm. He liked the feeling of the warm metal on his tummy.

Inside is very, very good

A few more weeks passed and now Smoky came out during the day to eat a little. One day, both the boy and the man were at home. The house was quiet, and, even though it was daytime, Smoky felt safe enough to go to his food bowl. As he walked through the den where the boy and the man were, the boy shouted his name. Surprising himself, Smoky didn't run for the kitchen but carried on walking. The boy jumped up and followed him. Smoky went to his dish and was eating when he noticed the lady was at the kitchen counter taking things out of a large paper bag. He looked up and stopped chewing. The lady asked, "What's wrong, Smoky?"

The weak winter sun peaked through the curtains, and Smoky saw the paper bag and remembered ... He had not always lived in the neighborhood, he had been born in a garage. His mom had soft grey fur, and she gave him baths and purred whenever he and his siblings nursed. A woman – like the lady in this house – would come out to the garage and talk to his mom and her kittens, scooping up Smoky and nuzzling him, giving him little scratches.

One night a man collected him and his siblings and put them in a paper bag which he placed in his car and drove for a while. Smoky and his brothers and sisters were scared, and meowed the whole way. The man eventually pulled over and dumped the bag of kttens in a driveway. Smoky ran for a hole under the house there. He didn't know what happened to his brothers and sisters as they all ran in different directions.

Thus it was that this neighborhood became Smoky's stamping ground. He would get into tussles with other cats; sometimes people would feed him, and one old lady began calling him Smoky because of his fur colour. He sort-of liked her but she ended up going away. and Smoky was busy being the sheriff until now. But here he didn't need to be the sheriff.

"Are you okay?" The lady bent down and stroked Smoky behind his ear. It was as if she knew the bag was scary. Smoky had a choice: he *could* run for the bedroom, but he looked at the lady, smiling gently, and the boy's eyes were big as the moon ... they waited to see what Smoky would do.

Smoky thought for a second and then turned back to his dish to finish his meal, then walked to the den. The man had made a fire in the fireplace; it was warm, and Smoky jumped onto the ottoman in front of the fire to give himself a long bath.

Smoky never discovered who the man and lady from so long ago were, but he knew these were okay-people – even the boy. He thought about his old neighborhood and way of life: the bougainvillea house and the leaking radiator, and the signs on the pole. As much as he liked his old neighborhood he thought maybe this house was better. He never knew about one particular sign on the pole that warned the neighborhood about the coyote who had been seen, and what relation that had to the fact that he never saw Juniper again, and neither did he know about the conversations between the lady and the man about why a grey cat with a black tail might be a nice addition to their family.

All Smoky knew was that it was nice to be in front of a fire with a full tummy ... he purred softly ... the first time he had since he was a kitten so very long ago ...

Tracy Abbott Cook, former writer for *The Tonight Show* starring Jay Leno; wife of Charlie, mother of Jack, and mother of Stella the cat and Marybelle the dog

15 Rules for people

There are many stories about people who adopt a cat while they are vacationing at the beach for a month, then leave her on the street to fend for herself when they depart. I've read tales about couples who adopt a cat and then, when a baby arrives, dump kitty at the nearest shelter because they have become completely focused on the infant. There are even reports of people moving away and leaving behind their pets in the empty house ... My heart breaks when I see those accounts. We can avoid being one of 'them' if we follow eight humane guidelines when adopting our feline companions.

- A cat is not a disposable item. When we adopt we must fully appreciate that this will be a 15-20 year commitment/relationship
- Adopt; don't buy in a pet store or from a breeder. 1.4million unwanted cats are euthanized each year. There are 70 million homeless animals in the US today. Two millions cats are homeless in the UK, and Japan euthanizes over 200,000 cats every year
- Absolutely spay or neuter your animal. Millions of homeless pets (cats and dogs) suffer and die because of the desperate struggle to survive. Do your part to help reduce the suffering population of homeless cats
- There will be costs associated with the care and feeding of kitty. Daily food and water, routine vet exams, shots, spaying/neutering, someone to feed and change the litter when you're away, and their old age comfort care. The costs are not huge, but set aside a fund to care for kitty
- Follow T S Eliot's guidelines for naming a cat (https://www.youtube.comwatch?v=TXkLgtusza4), giving her a suitable name that is respectful. Cats also appreciate middle names

- New members of the family – kids and dogs – should be carefully and kindly integrated into your cat's life. Don't make her take a back seat to new additions
- Foster love and trust with your feline; never punishment and fear. You have control of how your cat develops emotionally and physically
- The bond you will establish with your feline companion is beyond words, and one of the deepest you will ever experience. Your cat will reflect who *you* are

WHY SPAY OR NEUTER?

LET THE NUMBERS SPEAK FOR THEMSELVES:

An unspayed female and an unneutered male, having 2 litters per year, with 2.8 surviving kittens per litter, all of whom are left to breed, can produce exponentially:

1 year: 12
2 years: 67
3 years: 376
4 years: 2,107
5 years: 11,801
6 years: 66,088
7 years: 370,092
8 years: 2,072,514
9 years: 11,606,077

Created by Marlene Morris for *The One Minute Cat Manager*

One Minute Cat Manager summary

- Cats are people and family, too
- Adopt or rescue; don't purchase from a pet store or breeder
- Budget for your cat's care
- Give your cat a name of honor and respect
- Remember that your cat is your mirror: she will reflect your behavior ... do you like what you see?

Afterword

Let's start back in time ... When I met Kac Young we were both in our late twenties, and worked together for a television production company. Kac was bright and funny, and there wasn't a lot of that at this company. I remember wandering into Kac's office (probably to avoid working; possibly hungover) to tell her that I was having a REALLY hard time deciding whether to buy concert tickets for The Who, or Jackson Browne. I mean, they were both so cool! Kac, on the other hand, had the financial section of the newspaper in hand, and *she* was wondering if she should buy Procter and Gamble stock, or did IBM seem like a better long-term investment? I had no idea what language she was speaking, and I didn't ask. Then, possibly because of my lack of investments, and my decisions to spend my money on over-priced concert tickets (and souvenir t-shirts, if I'm being honest), I ended up spending some time couch-surfing with friends, while searching for a hovel in my price range.

Meanwhile, Kac bought her first house – an entire house – in the Hollywood Hills! And still, all I thought was: "Wow! She's so lucky!" And, it wasn't long after that that Kac got her first cat – Lucy. An adorable feline, who was so loving and friendly that she made me believe that I too could handle a furry roommate. And so, I got Boops, an angry gray Tabby who hated me. Truthfully, she hated everyone, and lived under my bed, high-tailing it out my front door every chance she got.

Usually, old Boops didn't get too far and I'd find her, but one day she seemed to be gone for good. I really panicked. My boyfriend at the time (and future second husband) wouldn't come home and help me, so I called the only person I knew who might agree to help with the cat search without asking for money, booze, or to borrow my pick-up truck, and that was the author of this book, Kac Young. Not only did she drive from the Hollywood Hills to Westwood – which, time-wise, is equivalent to about a hundred miles if you're driving in any other part of the country – she also arrived with a plan! I'm fuzzy on the details, but I seem to remember Kac drawing a grid of my neighborhood, having an open can of tuna for each of us, and then a lot of walking and 'here, kitty- kittying.' And, lo-and-behold, Boops was found!

So, what I'm trying to say is, there have been many times in my life when I could have learned an important lesson, or gathered helpful knowledge from my friend, Kac, but I was oblivious. I have more evidence of Kac's intellectual curiosity, as well as her ability to research and reason; as well as her generosity with her time and knowledge, if you need it. I mean, the woman has PhDs! More than one! And what I have is a new kitten, so I am going to use this book so that I have a shot at a healthy relationship with my new kitty.

If you have a cat, or are thinking about adopting one, I think you should join me. This is a wonderful book.

Lyla Oliver
Television producer and writer

Glossary

Alter - neuter (applies to either gender)

Boar cat - an old term for an unneutered male cat (tomcat)

Caregiver - person responsible for a pet cat or for a feral colony: these days the term 'owner' suffers from political incorrectness
Castrate - to remove the testes (neutering of male cats)
Cat collector - person who acquires and hoards great numbers of cats but does not provide proper care for them. They are unaware of their own shortcomings or the distress they cause to the cats they acquire. It is a form of obsessive compulsive disorder
Cat hoarder - another term, possibly a better one, for a cat collector

Desex - neuter (applies to either gender)
Domestic - an animal who has become adapted to people over many generations, and has a genetic predisposition to tameness

Entire - unneutered, undesexed, unaltered
Ex-feral - a feral cat who has been tamed and now lives as a pet

Feral - an ex-domestic cat who has reverted to being fully wild or is the wild-born offspring of stray cats (never known domesticity)
Feral-domestic hybrid - a misleading term that should really be changed. The definition a hybrid between a wild cat species and a domestic cat. Strictly speaking, it should be wild-domestic hybrid because a feral cat is not a wild species but a domestic cat gone wild
Full tom - unneutered male cat; this term is used to distinguish intact males from neutered males now that the term 'gib' has largely fallen into disuse

Gib - a castrated male cat; most people use the expression 'a neuter' instead
Guardian - another 'politically correct' term for a pet cat's owner or a feral cat's caregiver

Half-pedigree - a cat with one pedigree parent. In true terms, a half-pedigree cat is still a moggy, since a cat either is or is not a pedigree; there are no half-measures. Often used when selling accidentally-bred kittens as it's thought to sound more attractive than moggy. Some individuals deliberately breed half-pedigree cats for the pet market.
Household pet - in cat show terms, any cat who is not registered for breeding or exhibition in a breed category; may be random-bred, purebred (unregistered) or pedigree. There are household pet classes in many cat shows, but most are dominated by pedigree cats or pedigree lookalike cats; only a few shows have classes for genuine random-bred household pet cats
Hybrid - a cross between two different breeds: eg Persian and Himalayan (outcrossing) or two different subspecies: eg Siberian tiger and Bengal tiger (intra-specific hybrid), or two different species: eg lion and tiger (inter-specific hybrid)

Inbreeding - mating closely-related cats (sibling/sibling, mother/son, father/daughter) to strengthen desirable traits.
Intact - unneutered, undesexed, unaltered

Kitten - a young cat. Some cat regulatory bodies define a kitten as a cat below a particular age; this is for the purposes of cat show categories. The popular definition of when a kitten becomes an adult is based on when she reaches full size and sexual maturity (5-6 months of age)

Moggy (moggie) - mixed breed, crossbred or random-bred cat; one who is not pedigree or purebred
Moggy breeder - a misguided person who deliberately (irresponsibly) breeds random-bred kittens purely for private sale or to supply pet shops. Individuals such as this either do not believe in neutering or are simply trying to make money. Unlike the pedigree or purebred breeder, they are not attempting to develop or perpetuate a particular 'look.' Moggy breeders contribute greatly to the overpopulation problem
Mutt-cat - mixed breed, crossbred or random-bred cat, not pedigree or purebred

Neuter (noun) - a castrated tom cat or (less usually) a spayed female cat; vets often use the terms 'male neuter' and 'female neuter'
Neuter (verb) - to surgically render sterile; applies to males and females but is usually used as a euphemism for castration

Outbreeding - mating unrelated individuals to improve type or vigor
Outcrossing - mating a pedigree cat of one breed to a cat of a different breed/type in order to strengthen/improve the breed or introduce new traits

Pedigree - belonging to a particular breed and having a family tree registered with the breed regulatory body. A pedigree cat is not necessarily purebred since some breeds have allowable outcrosses: eg can be crossed to certain other breeds to improve characteristics. Some purebreds are bred for a subset of traits within a breed: eg for their color, and the term indicates that they have not been crossed to same-breed cats who are a different color.
Purebred - having only individuals of the same breed in the family tree; no outcrossings; not all purebreds have pedigrees: some purebreds are unregistered or the variety is not a recognized breed

Queen - an unspayed female cat. Also spelled quean

Ram cat - an old term for an unneutered male cat (tomcat)

Semi-feral - a non-domestic cat who lives in close proximity to people and is accustomed to human presence while remaining wild: eg a farm cat or barn cat
Spay (noun) - a spayed female cat; a female neuter. Also spelled spey
Spay (verb) - to remove the ovaries and womb (ovario-hysterectomy)
Sterilize - neuter (either gender)
Stray - a domestic (tame) cat with no home or owner

Tame - tamed in the cat's own lifetime, but born wild
Teaser tom - a vasectomized male cat: a teaser tom is sterile but behaves like a full tom and will spray, fight and mate
Tom - a male cat, particularly an uncastrated male cat; also called a 'full tom'

Unowned - an ownerless cat: a stray

Vasectomise - to snip the vas deferentia, but leave testes intact; the cat behaves like a full tom, but cannot make a female pregnant

Wild - often used to denote a feral cat; a wild cat is strictly a member of a non-domestic species: eg European Wildcat, Jungle cat, etc

Special thanks to Sarah Hartwell for allowing me to use her glossary in this book.

Index